PHYSICS IN THE LIFE SCIENCES

TO MAGGIE, HAMISH AND ANNA

Physics in the Life Sciences

GEORGE DUNCAN BSc, MSc, PhD

Reader in Biophysics
School of Biological Sciences
University of East Anglia
Norwich

SECOND EDITION

BLACKWELL SCIENTIFIC PUBLICATIONS

OXFORD LONDON

EDINBURGH BOSTON MELBOURNE

© 1975, 1990 by
Blackwell Scientific Publications
Editorial offices:
Osney Mead, Oxford OX2 0EL
8 John Street, London WC1N 2ES
23 Ainslie Place, Edinburgh EH3 6AJ
3 Cambridge Center, Suite 208
  Cambridge, Massachusetts 02142, USA
107 Barry Street, Carlton
  Victoria 3053, Australia

First published 1975
(with the title *Physics for Biologists*)
Second edition 1990

Set in Times by Times Graphics,
Singapore.
Printed and bound in Great Britain
by The Alden Press, Oxford

Library of Congress
Cataloging-in-Publication Data

Duncan, George, 1943–
  Physics in the life sciences/George
  Duncan. — 2d ed.
    p.   cm.
  Rev. ed. of: *Physics for biologists*.
  1975.
  Includes index.
  ISBN 0-632-01778-3
  1. Physics.   I. Duncan, George,
  1943–   Physics for biologists.
II. Title.
QC21.2.D86   1990
530′.02′4574—dc20

DISTRIBUTORS

Marston Book Services Ltd
PO Box 87
Oxford OX2 0DT
(*Orders:* Tel: 0865 791155
        Fax: 0865 791927
        Telex: 837515)

USA
Publishers' Business Services
PO Box 447
Brookline Village
Massachusetts 02147
(*Orders:* Tel: (617) 524 7678)

Canada
Oxford University Press
70 Wynford Drive
Don Mills
Ontario M3C 1J9
(*Orders:* Tel: (416) 441 2941)

Australia
Blackwell Scientific Publications
(Australia) Pty Ltd
107 Barry Street
Carlton, Victoria 3053
(*Orders:* Tel: (03) 347 0300)

British Library
Cataloguing in Publication Data

Duncan, George
  Physics in the life sciences. — 2nd
  ed
  1. Biophysics
  I. Title
  574.19′1

ISBN 0-632-01778-3

Contents

Preface to the Second Edition

The experimental basis of research in Medicine and Biology has radically changed over the past few years. The advent of Molecular Biological Techniques has led to a greater understanding of gene regulation and expression. This, in turn, has enabled new approaches to be taken both to investigate the molecular mechanisms underlying a wide range of diseases and to develop new drug technologies to control them. The introduction of a range of new physical techniques has, at the same time, permitted biologists and clinicians to study the molecular mechanisms underlying complex human physiological processes. For example, modern imaging techniques permit a study of molecular interactions and movements within individual cells and whole tissues and novel electrophysiological methods allow an investigation of the individual ionic pathways or channels through the cell membrane. Most of us working in the Life Sciences area have benefited from a knowledge of these techniques and one aim of this book is to provide an outline of the basic principles involved, which a wide range of readers from undergraduates to researchers may find useful.

However, I have a wider aim in writing this book and that is to inform teachers, pupils and those involved in planning science programmes in schools and universities that there is now a viable alternative to the conventional way of teaching the basic principles of physics. They will see, I hope, that nothing is lost in the rigour or, indeed, of the historical or broader perspectives of physics since biophysical investigations of the mechanisms of human physiology are as old as physics itself.

Preface to the First Edition

It is my firm belief that a successful 'Physics for Biologists' course can only be given by biologists or at least by those who are sympathetic towards biology. Undergraduates today are mercifully unwilling to put up with the Syrup of Figs attitude to this part of their education; they no longer accept their physics medicine with the vague promise that although it may not taste good now, it is surely going to have some beneficial effect in the not too distant future. They want to see the immediate relevance of the subject *now* and it is for these reasons that I have written this text.

I also hope that it will give a further insight into physics to those biology teachers who are giving a course and an insight into biology to those teachers in traditional Physics (or Natural Philosophy) departments who are given the onerous task of entertaining restive biologists. Most of all, however, it is to you restive biologists that this text is addressed. Ideally, it would take the form of a short explanation of the few basic principles of physics involved, followed by a long, a very long, reading list of original and review articles illustrating the biological applications of these principles. However, I realize that time is in short supply, that there are physical chemistry, organic chemistry, mathematics, statistics, computing and even biology lectures to attend as well, so I have here interwoven the principles and applications. At the end of each chapter I have listed some research articles and it is important that you should read at least one or two of those in a field that specially interests you, so that you can see how much more alive the subject *Biophysics* becomes on the battlefield itself.

As biologists we are continually faced with the situation where we have to describe a biological system in basic physical terms in order to learn more about the underlying physiological mechanisms. For example, Bennet-Clark and Lucey working in Edinburgh used hundreds of metres of high-speed photographic film in order to analyse the jump of the human flea. In this way they were able to time the jump and to measure the maximum height reached. When they then worked out the overall energy consumption for the jump they came to the startling conclusion that the flea's muscles simply could not supply this energy in the time available and this led them to propose an entirely new jumping mechanism. In the field of Phloem Translocation there is a similar problem because a relatively simple computation shows that conventional driving forces (e.g. osmotic pressure) are insufficient to transport water and solutes down the phloem at the rates that are normally found. This has stimulated a search for more unconventional mechanisms, which in turn has led to a reinvestigation of plant ultrastructure.

Both of these examples (along with many others) are set as problems and the answers, given in full, should be read as an integral part of the text. Problems of a revision nature are also set in the appropriate section and fully-worked answers are given at the back. The problems are presented in this way rather than as an indigestible lump at the end of each chapter so that problem solving can be seen as a valuable and integral part of a biophysics course rather than a chore to be sweated over in the tedium of the separate tutorial.

Acknowledgements

I wish to thank the faculty and students of the School of Biological Sciences at the University of East Anglia (UEA) for many discussions on all aspects of this text and, in particular, I would like to thank Drs D.J. Aidley, J.A. Bangham, S. Bassnett, T.J.C. Jacob, E.J.A. Lea, R.M. Warn and Professors R.W. Horne and E. Rojas for stimulating debate on specific issues. Drs G.R. Moore of the School of Chemistry, UEA and R.J. Ordidge of the Department of Physics, the University of Nottingham helped bring me up to date on several topics and Dr P.C. Croghan has given unfailing encouragement and penetrating, yet gentle criticisms throughout. Mr M. Williams also gave useful criticism at the proof stage. I also happily acknowledge the patient and expert assistance of the staff of Blackwell Scientific Publications. Finally, I owe an immense debt of gratitude to my wife Maggie, who not only produced the final draft manuscript in word-processed form, but ensured that my interest in, and enthusiasm for, this project has been maintained over the years.

1 Historical Perspective

Biophysics is not a product of the 20th century but has been in existence as a discipline for as long as physics itself. It is interesting to see that even 400 years ago, advances in physics were rapidly incorporated into the life sciences.

Galileo (1564–1642) is probably the best known Renaissance astronomer and physicist and he carried out many of his measurements with the aid of the pendulum. A contemporary of his, Sanctorius of Padua (1561–1636), introduced to medicine the first accurate measuring device, which he called the pulsilogium. The basis for this pulse-measuring instrument was, of course, the pendulum. Sanctorius built several of these (Fig. 1.1) but the principle of operation was the same. The frequency of oscillation of the pendulum — often a lead bullet at the end of a silk cord — was varied by altering the length of the cord until the frequency of the oscillation matched the pulse beat. Sanctorius

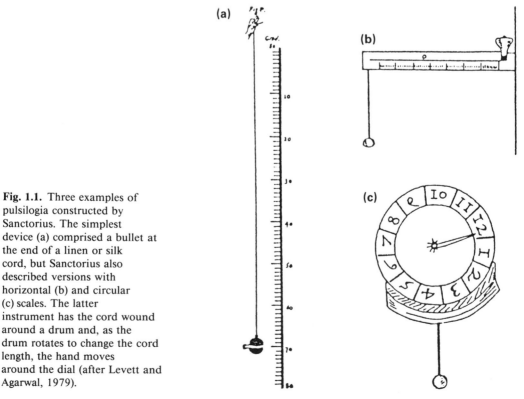

Fig. 1.1. Three examples of pulsilogia constructed by Sanctorius. The simplest device (a) comprised a bullet at the end of a linen or silk cord, but Sanctorius also described versions with horizontal (b) and circular (c) scales. The latter instrument has the cord wound around a drum and, as the drum rotates to change the cord length, the hand moves around the dial (after Levett and Agarwal, 1979).

established for the first time that the pulse was not universally constant, it not only varied from person to person, but he also found that he had to change the length of the synchronous pendulum according to the mood or excitement of his patient. It is interesting here that not only do we have the introduction of a 'null' method for determining pulse rate, but a measurement that is primarily in the time domain has been transformed to one in the length domain.

Following the Renaissance, the advance of physics was greatly accelerated by Sir Isaac Newton (1642–1727) who laid the foundations for both modern mechanics and optics (Chapters 2 and 8). One of his most ardent disciples was the Scottish physicist, Sir David Brewster (1781–1868), who, because of his scepticism of contemporary wave theories of light, was cuttingly described as the 'greatest 18th century physicist of his day'. Brewster's objections were based on the inability of the wave theory to explain the multiplicity of lines in the sun's absorption spectrum. Sir David was, however, quite content to apply the new wave mathematics where appropriate and his work concerning the interpretation of the optical diffraction patterns (see Section 8.12) from lens fibre cells represents an outstanding example of the application of physical and mathematical principles to biological problems. Normally the lens is perfectly transparent and there is no hint of a definite ultrastructure but, when an aqueous extract is made, long fibre-like cells can clearly be seen. Several contemporary scientists held the view that the fibres were merely an artefact of the method of preparation and so Brewster set out to resolve this by investigating the structure of the lens without recourse to an aqueous extract. He dissected a piece from the lens (in his case from edible cod) and placed it in a brass holder through which he could view a candle (Fig. 1.2). He found in fact that he could see very strong diffraction patterns on the horizontal plane which told him that there were regularly repeating structures in the lens. From the distance apart of the diffraction images he was able, by employing the new Fraunhoffer wave mathematics, to

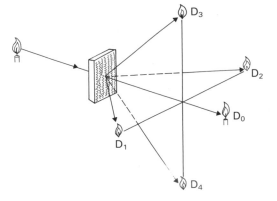

Fig. 1.2. Sir David Brewster's optical diffraction system. Brewster dissected out small pieces of lens and placed them in an anular holder. On viewing a candle through the section, he observed a series of diffraction patterns (D_1–D_4) and from the spacing between D_1 and D_2 calculated the fibre spacing within the lens. The value of 6 μm that he obtained is perfectly acceptable today. From the fact that he also observed vertical patterns (D_3 and D_4) Brewster concluded that there were regular protrusions on the surface of the fibres.

calculate the width of the regularly repeating units. His reported value of 6 μm would be acceptable to lens physiologists today. He also noted the widely spaced diffraction patterns in the vertical plane and suggested they arose from regular structures lying along the long axis of the fibre cell. On viewing sections of the lens with the aid of his very fine Ross microscope (Fig. 1.3) he was able to identify these structures as invaginations on the surface of the lens fibre cell. These 'ball and socket' joints were rediscovered in 1975 with the aid of the scanning electron microscope (Fig. 1.4).

Brewster's researches did not stop there, however, since he had his friend Whitworth (of 'screw' fame) make up a finely ruled grating where the separation of the lines corresponded to 6 μm. To his satisfaction, Brewster found the diffraction images were in the same place in the

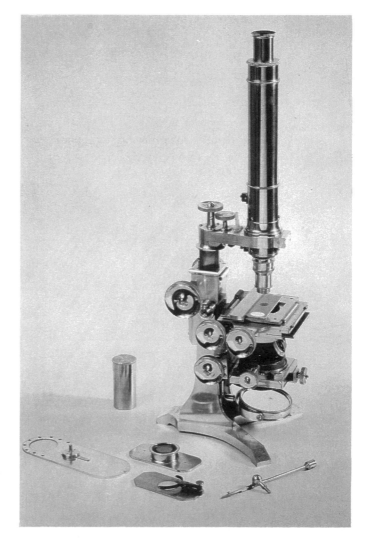

Fig. 1.3. Example of a fine Ross compound miroscope of the type used by Brewster and other scientists in the mid-19th century. The optical resolution of this microscope cannot be bettered by today's instruments (see Section 8.16). Below the object stage there is a secondary stage where a polarizing Nicol prism can be placed. This attachment allowed investigations of molecular arrangements for the first time (see Section 8.20). (From Turner 1981).

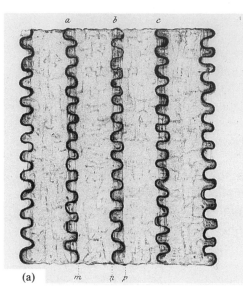

(a)

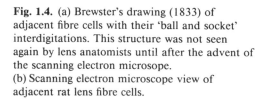

Fig. 1.4. (a) Brewster's drawing (1833) of adjacent fibre cells with their 'ball and socket' interdigitations. This structure was not seen again by lens anatomists until after the advent of the scanning electron microscope.
(b) Scanning electron microscope view of adjacent rat lens fibre cells.

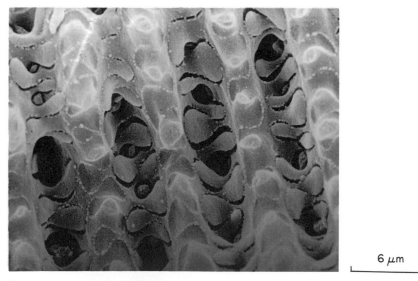

6 μm

(b)

horizontal plane. He also made an isinglass replica of the cut section of the lens and, on viewing the candle through the replica, he again observed diffraction patterns in both the horizontal and vertical plane, thus proving that they arose, not from any special peculiarities in the composition of the lens, but solely from regularities in the lens ultrastructure.

Brewster was thus able to relate abstract measurements from diffraction experiments to real structures in the lens and he used the most powerful mathematical techniques currently available to calculate the dimensions of the structures. He also employed the most accurate machinery of the day to build an analogue model to test his hypothesis. A scientific virtuoso performance hard to follow even in the latter half of this century.

The biophysical heirs of Newton, Young, Brewster, Helmholtz and a host of others, who have demonstrated that physics provides the key to unlock the mysteries of a whole range of biological problems, include Hodgkin and Huxley, and Crick and Watson. They worked in fields as diverse as neurophysiology and genetics but now these areas are brought together when molecular biologists and electrophysiologists combine to persuade cells such as oocytes to manufacture membrane channel proteins normally only found in neural tissues. In this way they have been able to elucidate not only the molecular structure of the channel protein but also the particular amino acid sequences that confer the physiological properties of the channel.

A complete understanding of the analysis that must be carried out before fully appreciating the detailed structure of any particular molecule or membrane channel would in fact be the subject of several postgraduate texts. The aim of this book however is, first, to provide an insight into how modern physical techniques can be applied to a range of biological problems and, second, to provide a firm theoretical basis from which the interested reader can launch into more advanced texts.

References

Levett J. & Agarwal G. (1979) The first man–machine interaction in medicine: the pulsilogium of Sanctorius. *Medical Instrumentation* **13**, 61–3.
Turner G.L'E. (1981) *Collecting Microscopes*. Cassell, London.

Further reading

Morrison-Law A.D. & Christie J.R.R. (1984) (eds) *Martyr of Science: Sir David Brewster 1781–1869.* HMSO, London.
Talbot J.H. (1970) *A Biographical History of Medicine.* Grune & Stratton Inc., New York.

2 Basic Physics: Mechanical Properties of Matter

2.1 Introduction

Although scientific articles are published in many different languages, scientists are increasingly employing the same system of units (Système International or SI) while making measurements and reporting the results of their experiments. For example, in French, German or English language articles on cardiac research, the unit of blood pressure is the kPa (kilopascal). The pascal is a measure of force per unit area and has dimensions mass $\times$ length^{-1} $\times$ time^{-2}. Details of the SI units commonly used in biology and medicine are given in Appendix 1.

2.2 Defined quantities: mass, length and time

The foundation of any physical science rests on the definition of the fundamental quantities, mass, length, and time (m, l, and t). The universally accepted standard of *mass* is at present the mass of a lump of platinum in Paris but the secondary definition based on the mass of the ^{12}C isotope of carbon (see Section 13.3 for a definition of *isotope*) will probably soon become the accepted standard. In the SI system the unit of mass is the kilogram and on the atomic mass scale, one kilogram is defined as the mass of 5.02×10^{25} atoms of ^{12}C. *Length* is now defined in terms of the wavelength (Chapter 8) of the orange line of krypton and the SI unit is the metre. *Time* is defined in terms of the vibration of the caesium atom and the SI unit is the second. The correct abbreviations of the many SI units are given in Appendix I.

2.3 Derived quantities: scalars and vectors

There are many physical quantities derived from the above fundamentals. The *velocity* (**v**) of a body is defined as the rate of change of position of the body with respect to time and it has dimensions $l\,t^{-1}$ and units m s^{-1}. Mass, length, and time are examples of *scalar* quantities, which means they have magnitude only, whereas velocity is an example of a *vector* quantity having both magnitude and direction.

Fig. 2.1. The vector **AB** has both magnitude and direction.

A   B

A vector (**AB** in Fig. 2.1) is usually represented by an arrow, the length of which gives a measure of the magnitude of the vector and the direction of the arrow indicates the direction of the vector. In this text vectors will be denoted by letters in **bold** type.

Vectors are subject to the laws of vector algebra. The rules for addition and subtraction are relatively simple (Figs 2.2 and 2.3) but those for the multiplication of two or more vectors are more complex since the outcome of the operation may be either a vector or a scalar quantity (see Appendix 2). When a scalar and a vector quantity are multiplied, then the result is always a vector. For example, the product of mass and velocity yields the vector quantity momentum with dimensions $ml\ t^{-1}$.

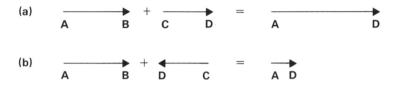

Fig. 2.2. (a) Vector addition. Vectors **AB** and **CD** in this case lie in the same direction and are said to be in phase. The resultant **AD** is also in phase with **AB**. In (b) the vectors **AB** and **DC** are antiparallel and are said to be 180° (or π radians) out of phase. The resultant **AD** is in phase with the vector of greatest magnitude. **AD** is in this case the result of subtracting **CD** from **AB**.

The addition law for vector algebra states that vectors are added from the tail of the first to the head of the last vector. For example, in Fig. 2.3, the vector **AC** is the net resultant of the two vectors **AB** and **BC**.

Fig. 2.3. The vectors **AB** and **BC** are ϕ radians out of alignment. The vector **AC** is the resultant.

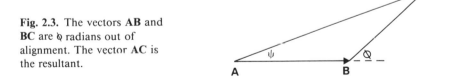

AB and **BC** are specified in magnitude and direction by the lengths AB and BC and the angle ϕ. The resultant **AC** is specified by the length AC and the angle ψ. AC and ψ can be obtained either by graphical or trigonometrical techniques. The vector equation is written

$$\mathbf{AB} + \mathbf{BC} = \mathbf{AC}. \tag{2.1}$$

When ϕ is zero, then the resultant **AC** lies along both **AB** and **BC**.

The rule for vector subtraction states that the vector to be subtracted should be rotated through 180° and the operation then treated as an addition (Fig. 2.2b).

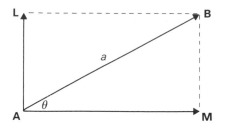

Fig. 2.4. Any vector can be resolved into two components at right angles. If the length of vector **AB** (the magnitude of a vector is often written **AB** or simply AB) is a, then the magnitude of **AL** and **AM**, the components of the original vector, are $a \sin \theta$ and $a \cos \theta$ respectively.

Any vector can be represented by two vectors at right angles to one another (Fig. 2.4). If the magnitude of **AB** is a, then the magnitudes of **AM** and **AL** are $a \cos \theta$ and $a \sin \theta$, respectively. **AM** and **AL** are said to be the components of **AB** and it is left as a simple exercise to show that when **AL** and **AM** are added tail to head, the resultant is **AB**. It is worth noting that θ is not unique; a vector can be resolved into components along any pair of axes.

Velocity is a vector quantity and the velocity of a body is defined as the rate of change of position(s) of the body with respect to time. In calculus notation (Ferrar, 1967, or any elementary calculus text) this is written as:

$$\mathbf{v} = \mathrm{d}s/\mathrm{d}t. \tag{2.2}$$

Acceleration is defined as rate of change of velocity with respect to time:

$$\mathbf{a} = \mathrm{d}v/\mathrm{d}t = \mathrm{d}^2 s/\mathrm{d}t^2. \tag{2.3}$$

Acceleration has dimensions $l\,t^{-2}$ and units m s^{-2}.

Momentum is another important vector quantity and is defined as the product of the mass of body and its velocity. Note that the product of a scalar quantity (such as mass) and a vector quantity (such as velocity) is a vector.

2.4 Mass and force: Newton's laws

Sir Isaac Newton was the great English scientist who formulated the basic laws governing the interaction of forces and masses. His three Laws of Motion form the basis for the science of mechanics and they are summarized below.

1 Every body continues in a state of rest or of uniform motion in a straight line unless it is compelled to change that state by an external force.

2 The rate of change of momentum of a body is equal to the resultant of all external forces exerted on the body.

3 When one body exerts a force on another, the second always exerts on the first a force, called the *reaction force*, which is equal in absolute magnitude, opposite in direction and has the same line of action.

Fig. 2.5. Summary of Newton's second law. A vector force **F** acting on a body of mass m produces an acceleration **a.**

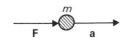

Newton's second law can be summarized in the equation

$$\mathbf{F} = \frac{\mathrm{d}}{\mathrm{d}t}(m\mathbf{v})\qquad\text{(2.4a)}$$

and as in most cases of biological interest the mass of the body remains constant while it is being acted on by the force (rockets in flight are an exception) equation (2.4a) simplifies to equation (2.4b),

$$\mathbf{F} = m\mathbf{a}.\qquad\text{(2.4b)}$$

In the SI system the unit of force is the *newton* which is the force required to give a mass of 1 kg an acceleration of 1 m s^{-2}.

Because different observers can describe the motion of the same body quite differently, a *frame of reference* must be introduced when describing the motion. For example, consider the hypothetical case of a passenger standing on frictionless roller skates in the corridor of a train that is accelerating away from a station (Fig. 2.6). His fellow passengers will note that he is accelerating backwards relative to them and so would say that he is experiencing a force. An observer standing at rest on the station would note that the roller-skated passenger remained more or less at rest and so would conclude that no force was acting on him.

Newton's first law cannot hold in both cases and it is in fact in the frame of reference fixed within the train that the law does not hold. There is no real backwards force on the roller-skated passenger, simply an apparent force. However, in order for Newton's first law to hold for an observer accelerating with the train, a force must be added to the passenger in a direction opposite to the acceleration of the moving frame. This force is sometimes called the *inertial* or *fictional* force.

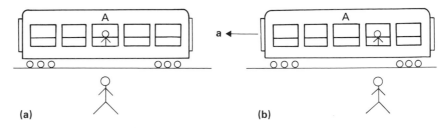

Fig. 2.6. Frames of reference (two views of the same event). (a) The roller-skated passenger is standing opposite window A when the train is at rest. (b) The train moves off with an acceleration **a**. Other passengers on the train will observe him moving backwards relative to them with an acceleration **a**, whereas an observer on the platform would say that he has scarcely moved during this time.

2.5 Gravitational and inertial forces

Newton also discovered the Law of Universal Gravitation which forms the basis of theoretical astronomy. It states that every body in the universe attracts every other body with a force **F** that is directly proportional to the product of their masses and inversely proportional to the square of the distance between them (Fig. 2.7). The magnitude of the force is given by the equation

$$\mathbf{F} = G\,\frac{m_1 m_2}{\mathbf{r}^2} \tag{2.5}$$

where G is the gravitational constant and is equal to 6.7×10^{-11} N m^2 kg^{-2}.

Gravitational forces are relatively weak compared to electrical (Chapter 10, p. 215) and nuclear (Chapter 13, p. 283) forces and are only significant when very large masses, e.g. planets, are involved.

The *weight* of a body is defined as the resultant gravitational force exerted on the body by all other bodies in the universe. The gravitational force of attraction on a body at the surface of the earth is such as to cause it to accelerate at about 9.8 m s^{-2} towards the centre of the earth. (The gravitational acceleration is usually denoted by the letter *g*.) The resultant force due to gravity on a mass of 1 kg is therefore 9.8 N or 1 kg weight. The weight of a body in fact varies from point to point on earth's surface as the gravitational acceleration varies due to inhomogeneities in the earth's composition and because the earth is not perfectly spherical. However, the mass of a body does not vary as it is determined by comparison with a standard mass on a balance.

The gravity-like nature of inertial forces is obvious in accelerating lifts (Fig. 2.8). (a) When the lift is accelerating upwards a person standing on a spring-balance weighing machine will note that his weight will have increased by an amount *ma* as the inertial force is in the same direction as the gravitational force. (b) Should the lift cable break the person will note that according to the machine he will have become weightless. However, if the person had been standing on an oversized

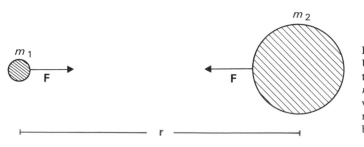

Fig. 2.7. Newton's Law of Universal Gravitation states that the force which m_1 exerts on m_2 is equal and opposite to that which m_2 exerts on m_1. The magnitude of each force is given by equation 2.5.

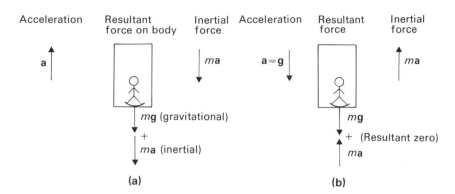

Fig. 2.8. (a) The inertial force adds to the gravitational force when the lift is accelerating upwards. (b) The person will appear weightless when the lift cable breaks.

chemical balance (one which compares his mass against a standard) his apparent mass would remain unchanged in the accelerated frames. In the case where the lift was moving with constant velocity, the weight as measured by the spring balance would remain unchanged because the frame of reference moving with the lift is not accelerating.

2.6 Bodies moving under the influence of a constant acceleration

Equation (2.3) can be used as the starting point for deriving several useful equations that are applicable when bodies move under a constant acceleration, e.g. a mass moving in a gravitational field.

Suppose we wish to know the velocity at any time t of a body which starts with an initial velocity $\mathbf{v}_0$ and is subject to a constant acceleration $\mathbf{a}$. Since the rate of change of velocity is constant, the graph describing velocity as a function of time will give a straight line of slope $\mathbf{a}$ (Fig. 2.9).

The equation representing the motion will have the form:

$$\mathbf{v}_t = \mathbf{v}_0 + \mathbf{a}t. \tag{2.6a}$$

Equation (2.6) is actually a vector equation and is only true when the vectors representing $\mathbf{v}_0$ and $\mathbf{a}$ are in the same direction (Fig. 2.10a).

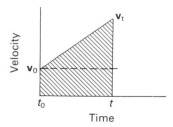

Fig. 2.9. Velocity–time graph for a body moving with constant acceleration.

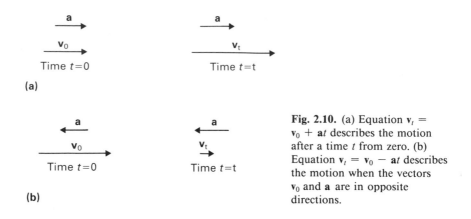

(a)

(b)

Fig. 2.10. (a) Equation $v_t = v_0 + at$ describes the motion after a time t from zero. (b) Equation $v_t = v_0 - at$ describes the motion when the vectors v_0 and a are in opposite directions.

When the vectors are in opposite directions (Fig. 2.10b) the equation takes the form

$$v_t = v_0 - at. \qquad (2.6b)$$

We know from elementary calculus that the area under the velocity–time graph represents the process of integration and is written $\int v\,dt$. Now, since velocity is the rate of change of position, the integral can be rewritten

$$\int \frac{ds}{dt}\,dt \quad \text{or} \quad \int ds.$$

Hence the area under the velocity–time graph (Fig 2.9) represents the total distance travelled by the body in time t. The area consists of a rectangle of area $v_0 t$ together with a triangle of area $\frac{1}{2}t(v_t - v_0)$. Hence, the displacement at time t is given by

$$s = v_0 t + \tfrac{1}{2}t(v_t - v_0). \qquad (2.7)$$

From equation (2.6a), we know that $(v_t - v_0) = at$ and substitution into equation (2.7) gives

$$s = v_0 t + \tfrac{1}{2}at^2. \qquad (2.8a)$$

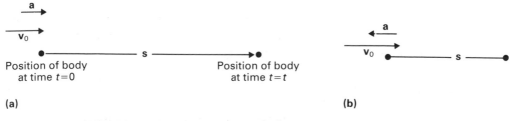

(a)

(b)

Fig. 2.11. (a) Equation $s = v_0 t + \frac{1}{2}at^2$ describes the motion.
(b) Equation $s = v_0 t - \frac{1}{2}at^2$ describes the motion.

Again, this is a vector equation and holds only when the velocity and acceleration vectors are in the same direction (Fig. 2.11).

When the velocity and acceleration vectors are in opposite directions,

$$\mathbf{s} = \mathbf{v}_0 t - \tfrac{1}{2}\mathbf{a}t^2. \tag{2.8b}$$

describes the motion (Fig. 2.11b).

It is now possible to produce a third set of equations by combining equations (2.6a) and (2.8a):

$$\mathbf{v}_t^2 = \mathbf{v}_0^2 + 2\mathbf{as}. \tag{2.9a}$$

and by combining (2.6b) and (2.8b) the equation is

$$\mathbf{v}_t^2 = \mathbf{v}_0^2 - 2\mathbf{as}. \tag{2.9b}$$

The three sets of equations (2.6, 2.8 and 2.9) are commonly referred to as *The Equations of Motion*.

Problem 2.1

A salmon jumps 5 m to clear a waterfall. With what initial velocity did it leave the water? Assume a negligible air resistance.

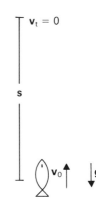

Fig. 2.12. See text for explanation.

Answer

Apply equation (2.9b): $\mathbf{s} = 5$ m, $\mathbf{v}_t = 0$, and $\mathbf{a} = 9.8$ m s^{-2}. This gives a result of approximately 10 m s^{-1} (or just over 20 mph) for the vertical component of the take-off velocity.

2.7 Work and energy

Forces, as we have already seen, are agents that are capable of accelerating masses but they can also be defined as the instruments by

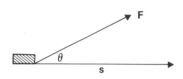

Fig. 2.13. As force **F** acts, the body actually moves along **s**, and work is done. This diagram could represent the work done in dragging a heavy object along a floor. Only the component **F** cos θ acts along **s** and so the work done is given by **Fs** cos θ. The vertical component **F** sin θ contributes nothing to the work done.

means of which the energy of a body is either increased or decreased. For example, a billiard ball lying at rest has its energy increased by hitting it with a billiard cue. The ball then gradually loses its energy again by frictional forces exerted on it by the table and also when it gives up some of its energy to other balls on the table.

When forces act on a body and the body moves, work is done by these forces. The work done as a body moves through a distance **s** is the product of the component of the force along **s** and the vector displacement **s** (Fig. 2.13), i.e.

$$\text{work} = \mathbf{F}\cos\theta\mathbf{s}. \tag{2.10}$$

The expression **F** cosθ**s** is the scalar product of the two vectors **F** and **s** and is written **F . s** (Appendix 2). Note that the displacement is a vector quantity, as the direction in which the body moves is obviously important.

Work and energy have the same dimensions ml^2 t^{-2} and both are scalar quantities. The SI unit of energy is the *joule* (J) which is defined as the work done when a force of one newton moves its point of application through a distance of one metre. There are two forms of energy:

1 *Potential energy* is accumulated in a system as a result of previous work being done on the system. It is energy stored in chemical bonds for example, or energy stored in a body that has been moved some distance above the surface of the earth. The stored energy is released once the bond is broken or the body is returned to the surface.

2 *Kinetic energy* is possessed by a moving body and is defined by the expression

$$\text{KE} = \tfrac{1}{2}m\mathbf{v}^2. \tag{2.11}$$

Consider the salmon problem again (Fig. 2.14). Equation (2.9b) gives

$$\mathbf{v}_t^2 - \mathbf{v}_0^2 = 2\mathbf{gs}$$

and multiplying both sides by $\tfrac{1}{2}$ m and rearranging

$$\tfrac{1}{2}\text{m }\mathbf{v}_0^2 - \tfrac{1}{2}\text{m }\mathbf{v}_t^2 = m\mathbf{gs}.$$

Now the work done against gravity by the salmon in jumping a distance

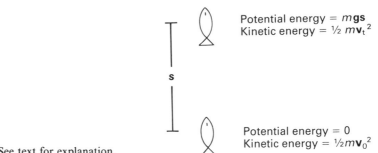

Potential energy = $m\mathbf{g}\mathbf{s}$
Kinetic energy = $\frac{1}{2}\,m\mathbf{v}_t{}^2$

s

Potential energy = 0
Kinetic energy = $\frac{1}{2}m\mathbf{v}_0{}^2$

Fig. 2.14. See text for explanation.

s is $m\mathbf{g}\mathbf{s}$ which is the increase in potential energy and this is equal to the change in the salmon's kinetic energy. This illustrates the conservation of energy theorem, as the potential energy gained equals the kinetic energy lost. If there had been frictional forces present, e.g. air resistance, then this would not be true as part of the initial kinetic energy of the salmon would be used to overcome these forces. This energy would be lost to the body and so would not be converted into potential energy.

The salmon moving solely under the influence of gravity is said to represent a *conservative* system as the work done is entirely accounted for by the change in potential energy of the salmon. When significant dissipative forces are present (such as those due to air resistance) this is no longer true and energy will be lost in the form of heat, for example. Such systems are said to be *non-conservative* and a book pushed at constant velocity along a table-top (Fig. 2.15) is an example of such a system. Work is done on the book by the force $\mathbf{F}_p$ and yet the potential and kinetic energies of the book remain constant. Four forces act on the system: $\mathbf{F}_p$, the force moving the book, $\mathbf{F}_r$, the frictional force which is proportional to the velocity; $m\mathbf{g}$, the weight of the book and $\mathbf{R}$, the reaction force. Energy is lost to the system mainly in the form of heat and the temperature of both the book and the table-top will increase.

Heat is a form of energy. The kinetic energy of the molecules making up a body determines the heat energy of the body and in the SI system the unit of heat energy is the joule (J). However, a unit sometimes used is the *calorie* (cal) and this is the heat required to raise the temperature

Fig. 2.15. A book pushed along a table-top moves with constant velocity under the action of four forces. It is in fact found in many frictional interactions that the force $\mathbf{F}_r$ is proportional to the relative velocities of the interfaces (**v**).

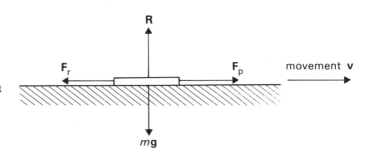

of 1 g of water by 1°C. Experimentally it has been found that

   1 cal = 4.2 J.

The kilocalorie is also much used, especially in metabolic rates which are usually quoted in kcal day^{-1}.

   Power is the rate at which energy is expended and it has units of force × distance × time^{-1}. If the instantaneous velocity **v** of a body moving under the influence of a force **F** is known, the instantaneous power developed P is given by

$$P = \mathbf{F} \cdot \mathbf{v}.$$ (2.12a)

The average power developed over a time is given by

$$P_{av} = \frac{1}{t} \int_0^t \mathbf{F} \cdot \mathbf{v} \, dt.$$ (2.12b)

The basic unit of power is the *watt* and

   1 W = 1 J s^{-1}.

   The inefficiency of man as a mechanical device is shown by the fact that while his basal metabolic rate is approximately 90 W, he can only develop about 50 W when working over long periods. When suitably stimulated, however, he can develop short surges of up to 250 W.

   The efficiency of movement of animals and man can be directly compared (Fig. 2.16) simply by measuring energy intake from food (kJ), distance travelled (km) and mass of the animal (kg). Migratory birds, however, consume fat during long flights and their efficiencies have to be estimated in wind-tunnel experiments where O_2 consumption and CO_2 output is measured.

   Man is a relatively efficient mover compared with most birds on an energy cost/km basis. However, some migratory birds, such as ducks and gulls can cruise at speeds of 30–80 m s^{-1} and many can fly for tens of hours without feeding. This means that several species of birds can travel 700–1500 km without stopping and this is far outside the range of endurance of all walking and running animals. It is interesting that on a weight basis, the salmon, another migratory species, is by far the most efficient traveller.

Problem 2.2

A footballer of mass 70 kg is running with the ball at a speed of about 5 m s^{-1}, i.e. he has potential energy in the form of glycogen and ATP and the more readily usable ATP is being converted into kinetic energy.
1   Find his kinetic energy and the power developed when he is stopped within 1 s in a tackle.

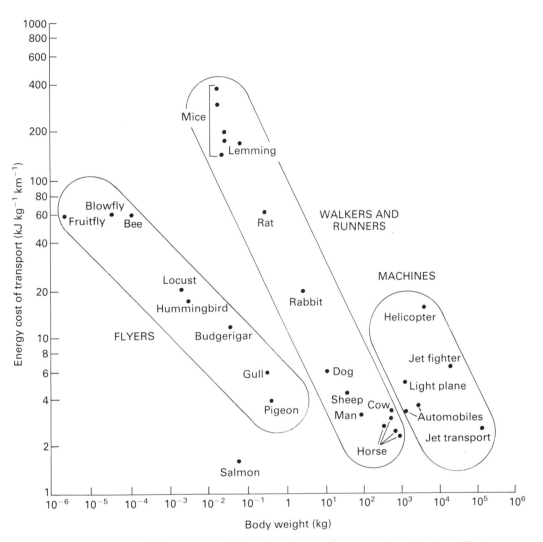

Fig. 2.16. The energy costs of various forms of movement as a function of body mass. Note that bird flight, when compared with other types of animal locomotion, is more economical than walking or running. Large flying birds can travel further for each kJ per unit mass than a light plane or jet fighter. The young salmon's performance shows, however, that a fish can travel more economically than any other kind of animal (after Tucker, 1969.)

2   What is his deceleration (assumed constant) during this time, i.e. his **g** value, and how does it compare with the 40 **g** said to be imparted by a professional boxer's gloved hand to a stationary non-elastic target?

2.8   Friction

When frictional forces operate on a body the system is non-conservative. For example consider a book being pushed along a table with constant velocity **v** (Fig. 2.15). The body is in fact in equilibrium under the action of four forces: the weight of the body $m\mathbf{g}$, the reaction force **R**, the force from the hand $\mathbf{F_p}$, and the frictional force between the book and the table top $\mathbf{F_r}$. The equilibrium is described by two sets of equations

$$\mathbf{R} = m\mathbf{g} \text{ (action and reaction system)}$$

and

$$\mathbf{F_r} = \mathbf{F_p}.$$

It is found experimentally that these two equations are further related by a *coefficient of friction* μ_f where

$$\mu_f = \frac{\mathbf{F_r}}{\mathbf{R}}. \tag{2.13}$$

It is also found experimentally that μ_f is relatively independent of the velocity of the body (in this case the book) and of the area of the body in contact with the surface over which it is moving. μ_f depends in fact on the textures of the two surfaces.

Frictional forces also operate at the molecular level when molecules flow past one another and the resultant interaction gives rise to the viscous properties of fluids. (Chapter 5, p. 74)

2.9 Circular motion

Consider a body of mass m travelling round a circular path (Fig. 2.17a) of radius **r**, with a tangential velocity **v**. The time to go round the circle is called the *period* of the motion (T). As the length of circumference of the circle is $2\pi\mathbf{r}$, then

$$\mathbf{v} = \frac{2\pi\mathbf{r}}{T}. \tag{2.14}$$

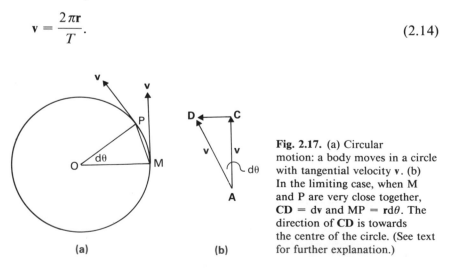

Fig. 2.17. (a) Circular motion: a body moves in a circle with tangential velocity **v**. (b) In the limiting case, when M and P are very close together, $\mathbf{CD} = d\mathbf{v}$ and $\mathbf{MP} = \mathbf{r}d\theta$. The direction of **CD** is towards the centre of the circle. (See text for further explanation.)

(a)   (b)

The frequency f is the number of cycles completed per second (hertz, Hz, is SI) and is the reciprocal of the period. Hence

$$f = \frac{\mathbf{v}}{2\pi\mathbf{r}}. \tag{2.15}$$

The angular velocity (ω) is the number of radians swept out by the radius in 1 s, i.e.

$$\omega = d\theta/dt$$

and as 2π radians are swept out in T s,

$$\omega = \frac{2\pi}{T} = 2\pi f. \tag{2.16}$$

Hence

$$\omega = \mathbf{v}/\mathbf{r}. \tag{2.17}$$

As the velocity vector of the mass is constantly changing direction, the mass will experience accelerating forces. To work out the acceleration, consider the velocities at the two positions M and P close together on the circle (Fig. 2.17). **AC** and **AD** represent the velocity vectors at these positions and the change in velocity is represented by the vector **CD**. From triangles OMP and ACD,

$$\frac{d\mathbf{v}}{\mathbf{v}} = \frac{\mathbf{r}d\theta}{\mathbf{r}}. \tag{2.18}$$

Hence the acceleration of mass is given by

$$\mathbf{a} = \frac{d\mathbf{v}}{dt} = \mathbf{v}d\theta/dt.$$
$$= \mathbf{v}\omega \text{ or } \mathbf{v}^2/\mathbf{r} \text{ or } \omega^2\mathbf{r}. \tag{2.19}$$

As the acceleration of the mass is directed towards the centre of the circle (Fig. 2.17b), the inertial or *centrifugal force* of magnitude $m\mathbf{v}^2/\mathbf{r}$ will be directed *away from the centre*.

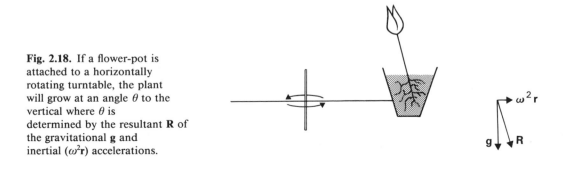

Fig. 2.18. If a flower-pot is attached to a horizontally rotating turntable, the plant will grow at an angle θ to the vertical where θ is determined by the resultant **R** of the gravitational **g** and inertial ($\omega^2\mathbf{r}$) accelerations.

As the effects of gravitational and inertial forces are indistinguishable, plant roots will grow away from the centre when the plant is set to grow on a rapidly rotating turntable, just as they normally grow towards the centre of the earth. The angle at which they grow depends on the resultant of the gravitational and inertial forces (Fig. 2.18).

Molecules in a rapidly spinning centrifuge tube experience accelerations that are several times g and so sedimentation is speeded up.

2.10 Rotating vector diagrams

Circular motion and wave motion (Chapters 7 and 8) are examples of *periodic motions* which can be represented by one or more rotating vectors. Light and sound waves can be represented in this way (Fig. 2.19). The magnitude of the vector represents the amplitude of the motion and the time taken for the vector to complete one revolution represents the period of the motion. The interaction of two simple harmonic motions, e.g. the interaction of two sound (Chapter 7, p. 104) or light waves (Chapter 8, p. 140), can be represented by the resultant of the two vectors representing the individual motions.

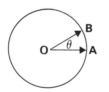

Fig. 2.19. Rotating vectors. The vector **OA** represents the motion at some arbitrary zero of time and **OB** represents the motion t seconds later. θ/t is the angular velocity of the motion and θ is the *phase* angle.

2.11 Mechanical equilibrium

In the preceding sections it has been possible to discuss the physics of moving bodies by treating them simply as point masses. However, in many problems, e.g. in animal mechanics, both the size and shape of a body are important, and also the point of action of any forces on the body. These problems can be solved by applying two basic laws of physics concerning the translational and rotational equilibrium of rigid bodies. It is also helpful to introduce the concept of the centre of gravity of a rigid body.

1 Translational equilibrium and centre of gravity

When a body is at translational equilibrium, the body either remains at rest or continues to move with a constant vector velocity. Under these circumstances, the vector sum of all the forces acting on the body is zero. The motion of a rigid body depends critically on where it is

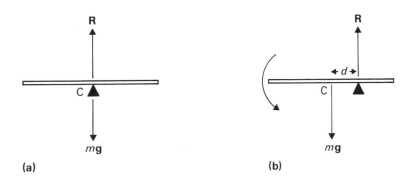

Fig. 2.20. (a) The homogeneous plank rests on a pivot at its centre and here it will be in equilibrium. (b) The fulcrum is towards the right and so the plank will swing downwards under the action of a force $m\mathbf{g}$ with moment $m\mathbf{g}d$. C is the centre of gravity of the plank.

pivoted. For example, if it is a uniform plank then it will remain in equilibrium, i.e. in balance, if placed on a support at its centre. There are two forces on the plank, its weight $m\mathbf{g}$ acting vertically downwards, and a reaction force (Newton's third law) at the fulcrum acting upwards. At the point of balance, these forces are equal and opposite in magnitude and direction. This implies that there is a point in a body through which the weight of the body appears to act and this point is called the *centre of gravity* (Fig. 2.20).

The centre of gravity of most bodies can be found simply by suspending the body from different points by means of a thread. Alexander (1968) has determined the centre of gravity of a locust by photographing a formalin-treated specimen. The centre of gravity is given by the intersection of the two lines AA' and BB' and the position in fact depends on the position of the locust's hind legs (Fig. 2.21).

If the object being considered is a human leg at rest (Fig. 2.22a), it is at equilibrium and principally three forces are acting: $\mathbf{F}_1$ is the reaction of the floor on the bottom of the foot, $\mathbf{F}_2$ is the resultant of the forces

Fig. 2.21. Outlines traced from photographs of a preserved locust, suspended by a thread. The centre of gravity is at the intersection of the line AA' and BB'. (From Alexander, 1968a. Reproduced by permission of Sidgwick & Jackson Ltd.)

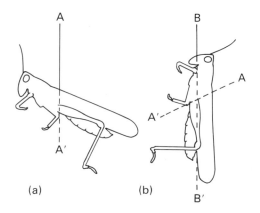

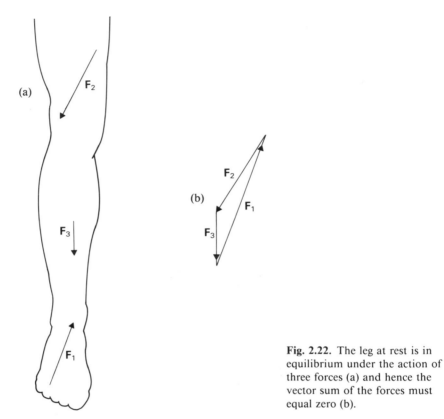

Fig. 2.22. The leg at rest is in equilibrium under the action of three forces (a) and hence the vector sum of the forces must equal zero (b).

from the rest of the body acting through the hip joint and surrounding muscles and F_3 is the gravitational force acting due to the mass of the leg. The condition for equilibrium requires that the vector sum of the forces is zero (Fig. 2.22b)

Note that as far as translational equilibrium is concerned, the point of application of the forces can be ignored.

2 Rotational equilibrium

If the uniform plank in Fig. 2.20 rests on a fulcrum at some distance d from the centre of gravity, it will rotate in a counter-clockwise direction. The greater the distance d, the greater will be the rotational force on the plank. This rotational force is termed the *torque* or *moment* and its magnitude is given by the product $m\mathbf{g}d$.

Fig. 2.23. Rotational equilibrium of a uniform plank balanced on a fulcrum distance d from the centre of gravity.

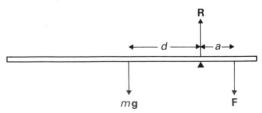

The plank can be brought into rotational equilibrium if another force is applied on the right side of the fulcrum such that its moment about the fulcrum is equal to $m$gd (Fig. 2.23), i.e.

$m$gd $=$ Fa.

In general, if an object is in *rotational equilibrium*, the algebraic sum of the moments of the forces acting on the body is zero. Note that the moments could be taken about any point in the plane containing the rod and the forces, but it is usually more convenient to choose a point on the rod. The point is often the fulcrum as the moment of the reaction force about this point is zero.

The condition for translational equilibrium also requires that

R $= m$g $+$ **F**.

The torque or moment can also be calculated even if the force is not at right angles to the rod.

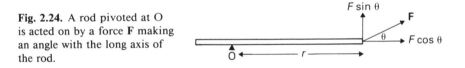

Fig. 2.24. A rod pivoted at O is acted on by a force **F** making an angle with the long axis of the rod.

The force (**F**) can be resolved into two components, **F** cos θ and **F** sin θ and only the latter of these has any turning effect on the rod. The moment of **F** about O is therefore r**F** sin θ.

Note that the moment has a positive sign when θ lies between 0° and 180° but is negative between 180° and 360°. These two conditions correspond to counter-clockwise and clockwise moments respectively.

2.12 The lever

When a plank is pivoted near one end at P it can be used as a lever (Fig. 2.25). There are four forces acting on the plank: the downward force F_d used to try to lift the heavy mass M, the reaction force **R** at the fulcrum, and the forces arising from the gravitational pull on the plank and stone. Three of the four forces will tend to rotate the plank about the fulcrum and the plank will be in equilibrium when the clockwise and counterclockwise moments are equal, i.e. when

$$\mathbf{F_d}\,(l - x) + m\mathbf{g}(l/2 - x) = M\mathbf{g}x$$

$$\mathbf{F_d} = \frac{M\mathbf{g}x + m\mathbf{g}(x - l/2)}{l - x}.$$

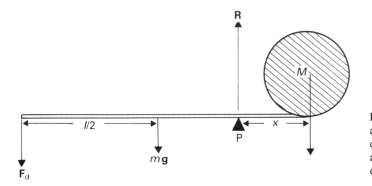

Fig. 2.25. A plank of mass m and length l pivoted near its end can be used to mechanical advantage. (See text for further explanation).

Hence $\mathbf{F}_d$ will be small if x is small and so the point of pivot should be as near the stone as possible. The ratio $M\mathbf{g}/\mathbf{F}_d$ is known as the *mechanical advantage* of the lever.

When the bicep muscles of the arm are used to lift a weight $M\mathbf{g}$ in the hand (Fig. 2.26) then the mechanical advantage of this lever system is given by

$$M\mathbf{g}/\mathbf{F}_b = \frac{l_b}{l_a}.$$

The force applied by the biceps is therefore greater than the weight to be lifted and the mechanical advantage <1.

Nutcrackers and jaws are other examples of levers. Among mammals there are considerable differences in the shape of the lower jaw (lever) and the relative sizes of the various jaw muscles (forces). These are correlated with the very different ways in which mammals with different feeding habits use their jaws.

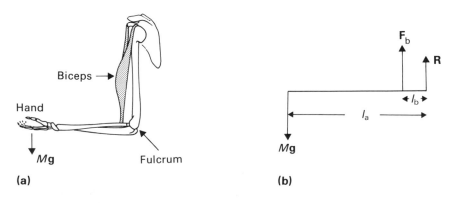

Fig. 2.26. Diagram of biceps muscles lifting a weight $M\mathbf{g}$ in the hand. In this system the mechanical advantage l_b/l_a is less than 1.

2.13   Motion of rigid bodies

The kinetic energy of a point mass moving with velocity $\mathbf{v}$ is $\frac{1}{2}m\mathbf{v}^2$. To compute the kinetic energy of a moving rigid body the shape of the body must be known, and also the axis about which the body is rotating. A simple example of the computation is provided by the insect's wing (Alexander, 1968). The wing is beating up and down about the axis indicated in Fig. 2.27.

Consider the wing to be divided into narrow strips cut parallel to the axis. The kinetic energy of a strip of mass m is given by

$$\text{KE} = \tfrac{1}{2}m\mathbf{v}^2 = \tfrac{1}{2}m\mathbf{r}^2\,\omega^2 \tag{2.20}$$

where $\mathbf{r}$ is the distance of the centre of gravity of the strip from the axis and ω is the angular velocity of the strips about the axis. The kinetic energy of the whole wing will be given by the sum of the kinetic energies of all the strips, i.e.

$$\begin{aligned}
\text{total KE} &= \Sigma\,\tfrac{1}{2}m\mathbf{r}^2\,\omega^2 \\
&= \tfrac{1}{2}\omega^2\,\Sigma\,m\mathbf{r}^2. \tag{2.21}
\end{aligned}$$

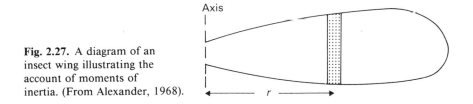

Fig. 2.27. A diagram of an insect wing illustrating the account of moments of inertia. (From Alexander, 1968).

Hence if each strip is carefully weighed and its distance from the axis of rotation also noted, the kinetic energy can be computed.

The sum $\Sigma m\mathbf{r}^2$ is called the *moment of inertia* of the wing and is denoted by $\boldsymbol{I}$, i.e. the rotational kinetic energy of any body is given by $\frac{1}{2}\boldsymbol{I}\omega^2$. Sotavolta has used the $\boldsymbol{I}$ value for the insect's wing to compute its energy balance in flight (Alexander, 1968).

2.14   Elasticity: Hooke's law

The important property of elastic materials is that they tend to return to their original shape when stretched. Elastic restoring forces are of great importance in biological systems as they provide a means of storing energy, e.g. kinetic energy imparted to the blood by the heart is stored in the elastic-walled arteries and serves to smooth the blood flow pattern.

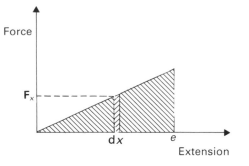

Fig. 2.28. Force extension graph for a perfectly elastic body. Work done when the force $\mathbf{F}_x$ moves its point of application through a small distance dx is $\mathbf{F}_x dx$ and this is given by the area of the small strip. The total work done in extending by an amount e is the total area under the graph.

Prefectly elastic bodies are said to obey Hooke's Law (Fig. 2.28) which states that the force **F** required to stretch the body is directly proportional to the extension x, i.e.

$$\mathbf{F} = kx \tag{2.22}$$

where k is the stiffness constant and has units N m^{-1}. The work done in extending the elastic body through a distance e is given by

$$\text{work} = \int_0^e \mathbf{F}\, dx \tag{2.23}$$

$$= \int_0^e kx\, dx \tag{2.24}$$

$$= \tfrac{1}{2} ke^2 \tag{2.25}$$

and this is the *elastic potential energy* stored in the stretched body. It is therefore the work which the stretched body can do when released.

2.15 Volume elasticity

The force per unit cross-sectional area on an elastic body fixed at one end is called the *tensile stress* (Fig. 2.29).

$$\text{tensile stress} = \mathbf{F}/A \tag{2.26}$$

and has units N m^{-2}.

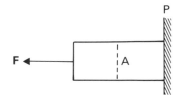

Fig. 2.29. An elastic body, attached at P and subject to a force **F**, experiences a tensile stress equal to F/A where A is the cross-sectional area of the body.

The term *strain* refers to the relative change in dimensions or shape of a body which is subject to stress (Fig. 2.30). The *tensile strain* is defined by the ratio e/l. which is dimensionless.

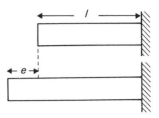

Fig. 2.30. An elastic body, originally length *l* is stretched by a small amount *e*, then the tensile strain is given by *e*/*l*.

When the force is parallel to the fixed edge (Fig. 2.31), the body undergoes *shear* deformation. The *shear stress* is the force divided by the area of the body and the *shear strain* is defined by the angle θ (radians).

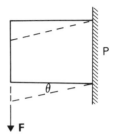

Fig. 2.31. when the force **F** is parallel to the fixed edge, the body undergoes shear.

2.16   Young's modulus

For perfectly elastic bodies, the ratio of tensile stress to strain is a constant Y, and is another way of expressing Hooke's Law.

$$\text{Young's modulus} = \frac{\text{tensile stress}}{\text{tensile strain}} = \frac{F/A}{e/l} = Y, \qquad (2.27)$$

i.e.

$$\mathbf{F} = \frac{YA}{l}e \qquad (2.28)$$

and comparing equations (2.22) and (2.28)

$$k = \frac{YA}{l}. \qquad (2.29)$$

Material	Young's modulus (N m^{-2})
Steel	2×10^{11}
Rubber	2×10^{6}
Resilin	1.7×10^{6}
Elastin	6×10^{5}

Table 2.1. Although all materials are to some extent elastic, this table shows that materials commonly thought of as elastic, e.g. rubber, have a low modulus of elasticity, i.e. they undergo relatively large deformations when stressed.

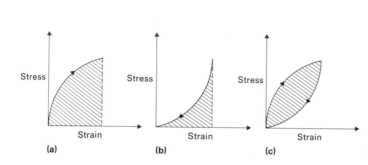

Fig. 2.32. Stress–strain curves for rubber. (a) The work done in stretching is the area under the curve. (b) The useful work recovered by the stretched body is released and is the area under the curve. (c) The total energy lost due to frictional interactions of the rubber molecules is given by the difference between the areas in (a) and (b), i.e. the hysteresis area.

The stress–strain curve for a piece of rubber does not follow a straight line (Fig. 2.32) and nor do the stretching and relaxing paths coincide. Energy is in fact lost in the cycle due to the frictional interaction of the molecules comprising the stretched material, and the amount is the area contained in the so-called *hysteresis* curve. Materials with very large hysteresis curves are very useful as vibration absorbers. During each vibrational cycle, the applied mechanical energy is largely dissipated in the form of heat, and so is not passed on through the system.

The elastic material called resilin found in many arthropods and serving different functions is remarkable in that it absorbs less than 5% of the vibrational energy even when forced to vibrate at 200 cycles per second (Hz). It is found at the base of insect's wings and when the wing is raised during flight the resilin is compressed and the stored energy is released to facilitate the down stroke. The scallop *Pecten* makes use of an elastic protein called abductin while swimming. The large adductor muscle closes the shell and compresses the ligaments which are then ready to force the shell open when the muscle relaxes. The scallop swims by opening and closing the shell three times per second and at this frequency abductin has a very low hysteresis loss (Fig. 2.33).

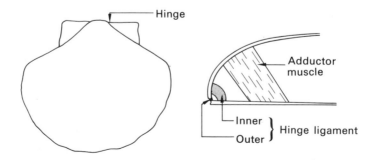

Fig. 2.33. Lateral view and diagrammatic transverse section of a scallop (*Pecten*), showing the position of elastic ligaments. (From Alexander, 1968). Reproduced by permission of Sidgwick & Jackson Ltd).

Problem 2.3 (from Bennet-Clark & Lucey, 1967)

Bennet-Clark and Lucey's work on the mechanism of the jump of the human flea is a model of its type in animal mechanics. They wished to know whether the insect's leg muscles alone could power the jump, or whether another source was involved. From careful measurements of the initial part of the jump, when the velocity of the insect was increasing, they concluded that the power output required was so great that it could not be provided by insect muscle. This led them to propose a novel jumping mechanism, powered by a material with unique elastic properties. Data were collected from film records of the normal velocity of the ballistic part of the jump of a flea.

1   The flea takes off with a vertical velocity of 1 m s^{-1} and reaches a height of 3.5×10^{-2} m. Show that the deceleration due to air resistance is half the gravitational deceleration.

2   The flea reached *from rest* its peak speed by 10^{-3} s. Show that the acceleration during this period was approximately 100 **g**. (This compares with the acceleration of 5 **g** on the body at lift-off in an *Apollo* moonshot.)

3   Given that the mass of the flea is 0.45×10^{-6} kg, find its kinetic energy after 10^{-3} s.

4   The maximum output of insect muscle is known from other sources to be 60 W kg^{-1} muscle. Hence, assuming that 20% of the flea weight is muscle, show that muscle alone cannot power the flight.

5   What is the energy store?

The answer lies in a resilin pad at the base of the hind leg (Fig. 2.34). Bennett-Clark and Lucey noted that the flea slowly bent its hind legs prior to a jump. In this way it would compress the resilin pads and some unknown trip mechanism could release the energy for the jump. All that remains is to show that the two pads each of volume 1.4×10^{-4} mm^3 can store sufficient energy.

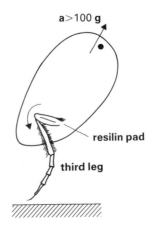

Fig. 2.34. Diagram showing the position of the resilin pad at the base of the flea's hind leg.

Assuming that resilin is perfectly elastic, the stored energy will be given by

$$PE = \tfrac{1}{2} ke^2 = \tfrac{1}{2} \frac{YA}{l} \cdot e^2.$$

The PE stored in 1 mm^3 which is compressed to zero will be given by

$$PE = \tfrac{1}{2} \times 1.7 \times 10^6 \times 10^{-9} \text{ (using } Y \text{ value given in Table 2.1)}$$

$$= 0.9 \times 10^{-3} \text{ J mm}^{-3}.$$

(Note: this is only an order of magnitude calculation as the strain is not small.)

The total stored energy from 2.8×10^{-4} mm^3 of pad is therefore 2.5×10^{-7} J per flea, which is just sufficient to power the jump.

2.17 Mechanics of the trunk

Painful strains and sprains of the lower back are common and in extreme cases herniation of the nucleus of an intervertebral disc (slipped disc) can occur leading to disablement.

The muscles of the back required during lifting are the erector spinae group. They are anatomically complex but for most purposes can be considered to act symmetrically through a line approximately 50 mm behind the centroids of the vertebral bodies (Fig. 2.35). The extension of these muscles induces a compression of the vertebral column and the compressive stresses involved will have units N m^{-2}. Experiments on cadavers have shown that damage can occur with stresses in the region of 4.6×10^6 N m^{-2} in males below the age of 60, while the threshold for damage is reduced to below 3×10^6 N m^{-2} above this age. The vertebral column is structured roughly in proportion to the weights of the body parts above the level concerned (Fig. 2.36) so that compressive stresses of approximately 0.3×10^6 N m^{-2} are encountered throughout the column in a relaxed, erect posture. These stresses increase dramatically

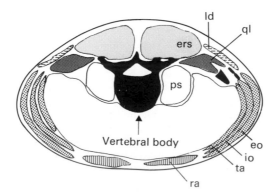

Fig. 2.35. Cross-section through the trunk in the mid-lumber region. Erector spinae (ers), latissimus dorsi (ld), psoas (ps), quadratus lumborum (ql), external oblique (eo), internal oblique (io), transversus abdominus (ta) and rectus abdominus (ra) muscles are shown. The first set play the major role when heavy weights are lifted.

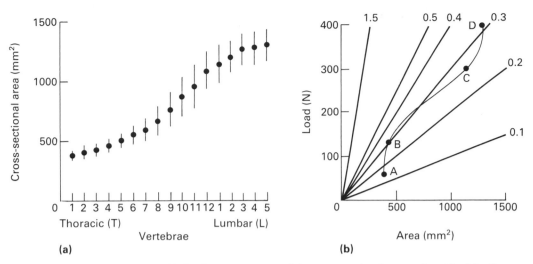

(a)

(b)

Fig. 2.36. (a) Skeletal measurements of the cross-sectional areas of vertebral bodies from 1st thoracic to 5th lumbar levels. (b) Approximate loadings on the vertebral column in a relaxed posture due to the weight of body parts above the level concerned. The lines of constant stress are numbered in N mm^{-2}. Points A, B, C and D refer to levels T_1, T_4, T_{12} and L_5 respectively. Note that compressive stresses of approximately 0.3 N mm^{-2} (0.3 × 10^6 Pa) are expected in the erect posture (after Grieve & Pheasant, 1982).

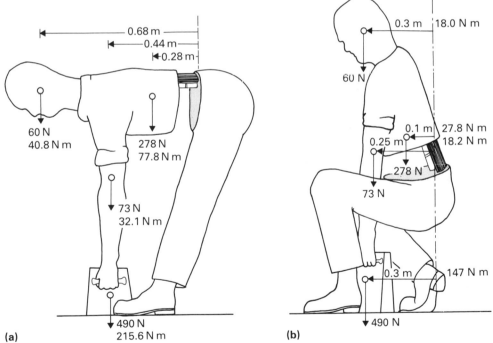

(a)

(b)

Fig. 2.37. (a) Man depicted in the initial posture of a stoop-lift. Horizontal distances of the centroids of the head, upper limbs, upper trunk and load with respect to the lumbosacral joint are shown. The weights (in N) of the parts and the turning moments (in N m) about the lumbosacral joint are indicated. The total turning moments before and after supporting the load are 151 and 366 N m respectively. (b) Man depicted in the initial posture of a crouch lift. The quantities indicated correspond to those in (a). The total turning moments about the lumbosacral joint before and after supporting the load are 64 and 211 N m respectively (after Grieve & Pheasant, 1982).

during lifting, but the possibility of damaging the vertebral column can be minimized by using the correct lifting procedure.

The International Labour Organization (ILO) recommends the maximum weight to be lifted by the adult male should not exceed 490 N. If a straight-legged posture (stoop-lift) is adopted, the total turning moment due to the weight, head, arms and trunk amounts to 366 Nm (Fig. 2.37). If we assume rotational equilibrium, when in fact the forces exerted are minimal, then a similar torque has to be exerted by the muscles. The compressive force on the lumbrosacral disc is therefore 7.32 kN if it is assumed that the erector spinae act through a line 50 mm behind the centroids of the vertebral bodies (Fig. 2.35). Since the cross-sectional area of the disc is approximately $1.3 \times 10^{-3}\,m^2$, the resulting compressive strain is $5.4 \times 10^6\,Nm^{-2}$. If the crouch-lift is adopted, the equivalent compressive strain is reduced to $3.1 \times 10^6\,Nm^{-2}$ and is below the hazard limit for the working adult male (Fig. 2.38).

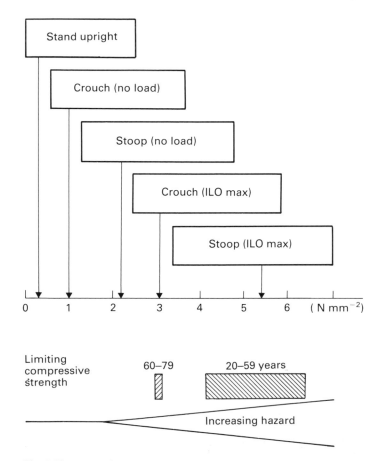

Fig. 2.38. Approximate compressive stresses on the vertebral column in the lumbosacral region, based on the data given in Figs 2.36 and 2.37, compared with limiting compressive strengths of cadaveric material (after Grieve & Pheasant, 1982).

2.18   Vibrations

Vibratory motion is common in nature and takes place, for example, when there exists a restoring force on a system whose magnitude is dependent on the displacement of the system from its equilibrium position.

For example, consider a spring of stiffness constant k, with a mass m at one end (Fig. 2.39). It is pulled down a distance x and released. If l_0 is the unstretched length of the spring, l the length at equilibrium with the mass attached, then

$$m\mathbf{g} = k(l - l_0) \tag{2.30}$$

When the spring is stretched by a further distance x, then the tension increases to $k(x + l - l_0)$. At equilibrium (i.e. before the mass is released).

$$\mathbf{F} + m\mathbf{g} = k(x + l - l_0)$$

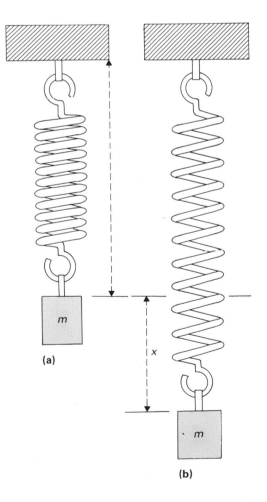

(a)

Fig. 2.39. (a) A spring of unstretched length l_0 is in equilibrium at length l when it has a mass m attached. (b) If it is stretched a further length x, then a force $\mathbf{F}$ is required to keep the mass in equilibrium.

(b)

where **F** is the force exerted to keep the system at equilibrium. However, when the mass is released (**F** = 0), then the system is no longer in equilibrium and the mass will experience an upward accelerating force. If the acceleration of the system is **a**, then

$$k(x + l - l_0) - m\mathbf{g} = m\mathbf{a}. \tag{2.31}$$

If x is taken as a positive quantity when the mass is below its equilibrium position, then the acceleration must be a negative quantity and can be written as

$$\mathbf{a} = -\mathrm{d}^2x/\mathrm{d}t^2 \ .$$

Substituting equation (2.30) into equation (2.31) then gives

$$kx = -m\,\mathrm{d}^2x/\mathrm{d}t^2.$$

The equation can also be written in the form

$$\frac{\mathrm{d}^2x}{\mathrm{d}t^2} + \left(\frac{k}{m}\right) x = 0. \tag{2.32}$$

The units of k/m are s^{-2} which are the units of frequency (f), or more correctly f^2. In vibratory problems, the solutions to the differential equation (2.32) are in terms of angular frequency (ω) where

$$\frac{\mathrm{d}^2x}{\mathrm{d}t^2} + \omega_\mathrm{n}^2\, x = 0. \tag{2.33}$$

and

$$\omega_\mathrm{n} = \sqrt{\frac{k}{m}}.$$

A solution to the above differential equation must satisfy the condition that $\mathrm{d}^2x/\mathrm{d}t^2$ always has the opposite sign to x. Common functions that satisfy this criterion are

$$x = A \sin(\omega_\mathrm{n}t + B) \tag{2.34a}$$

and

$$x = A \exp(i\omega_\mathrm{n}t + B) \tag{2.34b}$$

where A and B are constants and $i = \sqrt{-1}$.

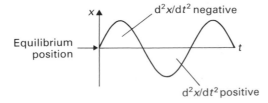

Fig. 2.40. Graph of position (x) against time (t) for a freely vibrating mass attached to a spring.

A graph of x against t will therefore have a sinusoidal form (Fig. 2.40).

2.19   Damping

Since vibratory processes are subject to energy losses due to frictional interactions, then the amplitude of the displacement from the equilibrium position will decrease with time.

In many systems, the frictional interactions (e.g. of the mass in air, or of a cylinder in a volume of oil) depends on the velocity of the body (Fig. 2.41) and the damping force ($\mathbf{F}_d$) will be given by

$$\mathbf{F}_d = -C\,\mathrm{d}x/\mathrm{d}t \tag{2.35}$$

where C is the damping constant.

In this case the equation of motion is

$$\frac{\mathrm{d}^2 x}{\mathrm{d}t^2} + \frac{C}{m}\frac{\mathrm{d}x}{\mathrm{d}t} + \frac{k}{m}x = 0 \tag{2.36a}$$

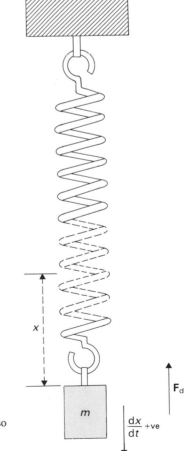

Fig. 2.41. The mass attached to the spring (Fig. 2.39) is now considered to make frictional interaction with the medium surrounding it and so will be acted on by an additional force ($\mathbf{F}_d$).

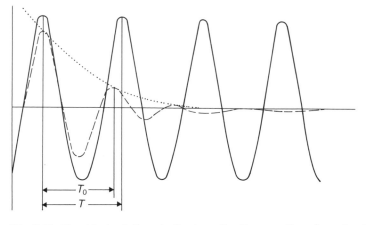

Fig. 2.42. Graphical solutions to the two vibration equations for natural (——) and damped (–––) vibrations. The sine wave (——) shows the undamped motion of period $T = 2\pi/\omega_n$. T_0 is the period of the damped motion (–––). The exponential decline of the amplitude of the damped motion is shown by the dotted curve (········).

or

$$\frac{\mathrm{d}^2 x}{\mathrm{d}t^2} + 2\,h\,\omega_n \frac{\mathrm{d}x}{\mathrm{d}t} + \omega_n^2 x = 0 \qquad (2.36\mathrm{b})$$

where $2h\,\omega_n = C/m$ and h is termed the damping factor.

The solution to the damping part of the equation alone, i.e.

$$\frac{\mathrm{d}^2 x}{\mathrm{d}t^2} + \frac{C}{m}\frac{\mathrm{d}x}{\mathrm{d}t} = 0$$

is exponential in form (try $x = \mathrm{e}^{-Ct/m}$) and in fact the solution to the whole equation consists of an exponentially damped sine wave of period $T_0 = 2\pi[(1 - h^2)^{1/2}]/\omega_n$. When h is small then the system is lightly damped and oscillates about the equilibrium position before coming to rest. When h is large then damping is heavy and the equilibrium position is only approached after a very long time (Fig. 2.42).

2.20 Resonance

When a damped system such as the one illustrated in Fig. 2.41 is also acted on by an oscillating force of magnitude $\mathbf{F}_0 \sin \omega t$, then the system will vibrate for as long as the force is applied.

The equation of motion becomes:

$$\frac{\mathrm{d}^2 x}{\mathrm{d}t^2} + \frac{C}{m}\frac{\mathrm{d}x}{\mathrm{d}t} + \frac{k}{m}x = \mathbf{F}_0 \sin \omega t. \qquad (2.37)$$

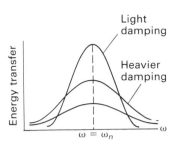

Fig. 2.43. Resonance curves for energy transfer between two vibrating systems.

The response of the system given, for example, by the ratio between the applied and response amplitudes depends on the ratio ω/ω_n. Energy can be transferred from one system (the oscillating force) to the other (vibrating mass) and the energy transfer is a maximum when $\omega = \omega_n$, the natural or undamped frequency, and this is also termed the resonant frequency. The shape of the energy transfer graph (Fig. 2.43) depends on the damping coefficient h and the curve is sharp around the resonance frequency when the damping is light.

Great care has to be taken in designing large structures, such as bridges and tall buildings, to ensure that resonance transfer of energy from the surroundings does not occur. High resonance transfer of energy from wind to the supporting structures of the Tacoma bridge was responsible for its destruction on 7 November 1940 (Fig. 2.44).

Fig. 2.44. The Tacoma bridge about to collapse due to resonance transfer of energy. (Photograph courtesy of UPI.)

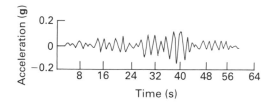

Fig. 2.45. Accelerogram obtained by the Instituto de Ingenieria in the soft-bed region of Mexico City during the 1985 earthquake. The amplitude reached 0.2 **g**, while in the lava beds to the south an amplitude reading of only 0.07 **g** was registered. Note that the period of the oscillation is about 2 s (after Flores *et al.*, 1987).

Resonant transfer of energy from the surrounding terrain to certain types of buildings has been suggested as the cause of the relatively selective destruction during the recent earthquake in Mexico City on 19 September 1985.

The destruction had in fact three remarkable features. First, there was a concentration of damage in soft terrain in the region that had formerly been a lake bed. Second, there was an alternating pattern within this region of high and low damage areas and, third, there was a high selectivity for buildings between five and fifteen storeys high. It appears that the earthquake set up an oscillating standing wave pattern in the soft terrain due to wave reflection at the surrounding firm ground. This explains both the alternating pattern of damage and the reason why it was greatest in the lake bed region. An accelerogram recorded in this region (Fig. 2.45) provides a clue as to why intermediate height buildings were most at risk. The trace shows the remarkable stability of the period of oscillation of about 2 s and this is in fact close to the natural resonating frequency of buildings between five and ten storeys high.

2.21   Resonance transfer in biological systems

The energy transfer curves in Fig. 2.43 have been obtained by solving the differential equations describing the systems. In practical terms it is not the energies of the systems which are measured, but rather the forces, velocities and amplitudes involved.

The ratio of the applied force and resulting velocity is termed the *impedance function* and the ratio of applied force to resulting acceleration the *transfer function*.

The impedance function data derived from man (Fig. 2.46) show expectedly that there is more than one characteristic frequency of response. The two most prominent resonant frequencies are at 5 and 12 Hz, although a Fourier analysis of the data (see Section 7.3) would have to be carried out to describe the system properly. In fact, the resonating man approximates well to a single simple mass–spring–damper system with a moderate damping factor of about 0.2.

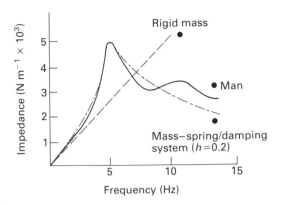

Fig. 2.46. Typical impedance function (ratio of force to velocity at driving point) curves for three mechanical systems. The rigid mass, attached to a spring, simply follows the vibrating force in a linear manner. In a damped system, however, resonance transfer occurs at specific frequencies.

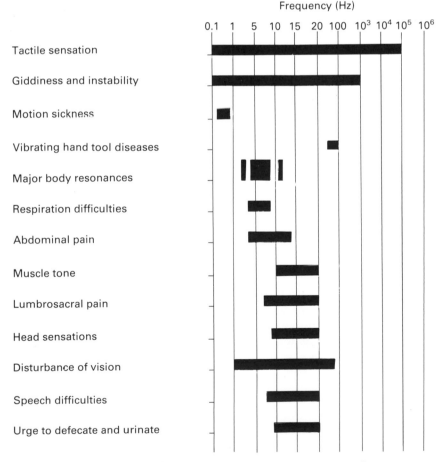

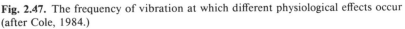

Fig. 2.47. The frequency of vibration at which different physiological effects occur (after Cole, 1984.)

The physiological, psychological and long-term medical effects of exposure to forced vibrations have been studied in some detail and the results are summarized in Fig. 2.47. The major resonance at around 5 Hz (see Fig. 2.46) arise from the respiratory system. Higher frequency resonances (around 12 Hz) are set up probably in the spine and gastrointestinal systems and drivers of agricultural tractors and cross-country vehicles, for example, have an increased risk of health problems in these areas because of the vibrations set up while driving over rutted fields. The mechanisms responsible for the health hazards associated with the long-term use of vibrating hand tools (white finger syndrome) are not understood but are associated with high frequencies of the order of 100 Hz. Prolonged exposure to loud noise can induce deafness. Major body resonances are not involved here, but rather the finely tuned resonances associated with small elements of the auditory system (Section 7.6).

Supplementary problems

2.4   A bush baby accelerates from rest to a take-off velocity of 7.0 m s^{-1} in 0.05 s. The muscles which provide the power for the jump constitute about 10% of the total mass of the animal. Estimate the total work done and the power expended by unit mass of jumping muscles.

2.5   From physiological experiments in the Antarctic, it has been shown that a daily diet of 20×10^6 J per dog will keep a team of huskies at constant weight while pulling sledges. Use of a strain gauge in the main trace to measure the total pull exerted by nine dogs gave an average value of 795 N while the dogs were walking at 1.55 m s^{-1}. The dogs normally work a seven hour day. (a) What is the rate of work (useful power output) of each dog while pulling? (b) What percentage of the energy intake appears as work done on the sledge by each dog?

2.6   A plant fixed to a turntable is rotating in a circle of radius 1 m. The tangential velocity of the plant is 3.2 m s^{-1}. (a) What is the magnitude and direction of the plant's acceleration? (b) Compute the angle which the growing roots make with the vertical (take **g** = 10 m s^{-2}).

2.7   A man is lifting a mass of 2 kg, as shown diagrammatically below:

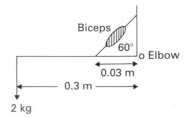

If the forearm (assumed uniform) weighs 0.5 kg, what tension is required in the muscle to maintain the position as shown? What is the force acting at the elbow joint?

2.8   The stress–strain graphs for various biological materials are shown below:

(a) What do the graphs tell us about the elastic properties of each of the materials? (b) Calculate Young's modulus for each material.

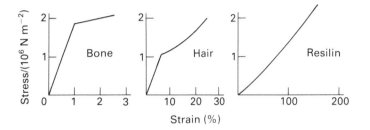

References

Alexander, R.M. (1968) *Animal Mechanics*. Sidgwick & Jackson, London.

Bennet-Clark, H.C. & Lucey, E.C.A. (1967) The jump of the flea; a study of energetics and a model of the mechanism. *Journal of Experimental Biology*, **47**, 59–76.

Cole, S. (1984) Vibration and linear acceleration. *In The Body at Work* (ed. W.T. Singleton). Cambridge University Press, Cambridge.

Ferrar, W.L. (1967) *Calculus for Beginners*. Clarendon Press, Oxford.

Flores, J., Novaro, O. & Seligman, T.H. (1987) Possible resonance effect in the distribution of earthquake damage in Mexico City. *Nature*, **326**, 783–5.

Grieve, D. & Pheasant, S. (1982) Biomechanics. *In The Body at Work* (ed. W.T. Singleton). Cambridge University Press, Cambridge.

Tucker, V.A. (1969) The energetics of bird flight. *Scientific American* (May), 70–7.

Further reading

Alexander, R.M. (1971) *Size and Shape*. Arnold, London.

Burns, D.M. & MacDonald, S.G.G. (1970) *Physics for Biology and Pre-Medical Students*. Addison-Wesley, Reading, Mass.

Jarman, M. (1970) *Examples in Quantitative Zoology*. Arnold, London.

Sears, F.W. & Zemansky, M.W. (1964) *University Physics*. Addison-Wesley, Reading, Mass.

3 Temperature and Heat

3.1 Temperature and thermal equilibrium

We have seen that in order to describe the motions of particles and rigid bodies, three fundamental quantities are required, namely mass, length, and time. These three quantities are sufficient to describe isolated bodies, but they have to be supplemented when bodies come into contact, and also when radiation falls on an isolated body. For example if we touch another body we say it feels either *hot* or *cold* relative to ourselves. In order to describe this feeling further, an additional fundamental quantity called *temperature* is invoked.

Two systems in thermal contact are said to be in *thermal equilibrium* when the temperatures of both bodies are equal. If two systems in contact are not in equilibrium then heat energy will flow from the hot body to the cooler one until the temperatures are equal. A system can be insulated from those around it so that there is no thermal interaction. Such a system is said to be in *adiabatic isolation*.

Temperature measurements are made quantitative by the introduction of a temperature scale and the two systems most widely used are the Centigrade and Kelvin scales. In the former, the melting point of pure ice at one atmosphere pressure is taken as the arbitrary zero of temperature (0°C) while the boiling point of pure water at one atmosphere pressure is taken as 100°C. In the latter scale the melting point of ice is 273 degrees Kelvin (273 K) and the boiling point of water is 373 K. Note that conventionally the expression degrees Kelvin is shortened to Kelvin.

As well as being defined in terms of the thermal equilibrium of bodies, temperature can also be defined in terms of the kinetic energy content of the atoms or molecules comprising the body (internal energy of the body). The kinetic energy of a gas atom, for example, depends solely on the temperature of the gas and from the Kinetic Theory of Gases (Sears & Zemansky, 1964) the average kinetic energy per atom is given by

$$\text{KE} = \frac{3}{2}kT \tag{3.1}$$

where T is the temperature of the gas in K and k (the Boltzmann constant) is equal to 1.38×10^{-23} J K^{-1}. The absolute zero of temperat-

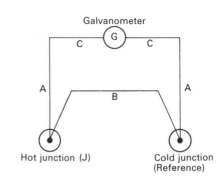

Fig. 3.1 An emf (see Chapter 10) is developed between the hot and cold junctions of two dissimilar metals A and B. These are connected, via copper wires C, to a simple galvanometer G, and the deflection observed can readily be calibrated in terms of the temperature difference between J and the cold bath which is commonly held at 0°C by stirred, melting ice.

ure 0 K corresponds to an absence of atomic or molecular movement.

A *thermometer* is a device for measuring temperature and the most common type, a liquid in glass thermometer, makes use of the fact that liquids expand when heated. *Thermocouples* (Fig. 3.1) are widely used for temperature measurements in biological systems and in these a thermoelectric voltage, usually measured with a galvanometer, is produced at the junction of two dissimilar wires. The usefulness of the thermocouple arises because the area of the junction in contact with the system under investigation can be quite small and also because it is simple to obtain relatively accurate measurements.

3.2 Heat and the First Law of Thermodynamics

Joule was the first to demonstrate two ways of increasing the temperature of a body in a quantitative manner. He showed that the final temperature of a can of water does not depend on the manner in which the energy is delivered. Y units of heat energy from a flame produce the same temperature rise as Y units of mechanical energy dissipated during vigorous stirring.

Joule's conclusion is stated in thermodynamic terminology by the following equation, called the *First Law of Thermodynamics*,

$$\Delta E = \Delta Q + \Delta W \qquad (3.2)$$

where ΔE is the increase in internal energy of the water, ΔQ is the heat energy supplied to the water, and ΔW is the work done on the water. The temperature of the water rises because there has been an increase in internal (thermal) energy.

3.3 Heat capacity

When a body receives an amount of heat ΔQ then its temperature rises. The extent of the temperature change (ΔT) depends on the mass of the body m and its specific heat S.

Substance	Specific heat (J kg^{-1} deg^{-1})	Table 3.1. Specific heat.
Copper	390	
Ice	2110	
Water	4200	
Biological tissue	c. 3700	
Air	950 (STP)	

The specific heat for air is given at 0°C and at a constant pressure of 1 atmosphere (10^5 N m^{-2}), i.e. the specific heat is said to be quoted at STP (Standard Temperature and Pressure).

The *specific heat* of a body is defined as the heat required to raise the temperature of a unit mass of the body through 1 degree (1°C or 1 K). The specific heat S is therefore defined by the equation

$$\Delta Q = mS\Delta T. \tag{3.3}$$

Examples of the specific heat of some common substances are given in Table 3.1.

3.4   Heat and change of phase

All substances exist in one of three phases—solid, liquid, and gas and the change from one phase to another is marked by a change in the internal energy of the substance. For example, 1 kg of water vapour has a much higher internal energy than 1 kg of water at the same temperature so that a relatively large amount of heat is required to vaporize the water. This heat is called the *latent heat of vaporization* and for water it is approximately 2.5×10^6 J kg^{-1}.

Warm blooded animals make use of latent heat as one means of regulating their body temperature. When the hypothalamus detects an increase in blood temperature it signals the sweat glands to increase secretion. The vaporization of this sweat requires latent heat energy and this is supplied by the body which is therefore cooled and so the hypothalamus activity is depressed. This is a good example of a biological feedback control mechanism.

3.5   Thermal interaction with the environment

All bodies, whether located for example on a sunny Norfolk beach or in a spacecraft in the vastness of space, exchange heat energy with their environment. If we consider a body on the beach we see that it will be heated by energy radiated from the sun above and energy conducted from the sand beneath. It will probably also be cooled by winds (convected energy) from the sea. If the body is alive, it too will be producing heat energy from chemical reactions. As all living systems can only work within a small temperature range, special devices (fur, sweating, large ears, etc.) are developed in order to maintain a stable temperature. We shall now consider the mathematics involved in describing heat transfer.

3.6   Heat transfer by conduction

When a poker is held with its tip in glowing coals, then the handle becomes gradually hotter although it is not in direct contact with the heat source. There is no gross motion of the body during this heat transfer, which is said to take place by conduction.

It is an experimental fact that the rate of heat flow depends on two factors. First it is proportional to the cross-sectional area through which the transfer is taking place, and, second, it depends on the temperature gradient between the source and receiving parts of the body.

If we consider a sheet of material (Fig. 3.2) of cross-sectional area A and thickness ΔX, with face 1 maintained at a temperature T_1 and face 2 at a temperature T_2, then heat will flow from 1 to 2 when $T_1 > T_2$, i.e. thermal energy travels down a temperature gradient.

This is expressed mathematically by Fourier's heat equation

$$H = KA \frac{\Delta T}{\Delta X} \tag{3.4}$$

where H is the heat flux in watts, K is the thermal conductivity of the material in W m^{-1} deg^{-1} and $\Delta T/\Delta X$ is the temperature gradient.

Equation (3.4) has the form

flow = constant $\times$ driving force

and in this case the driving force on the heat flow is the temperature gradient. A similar equation can be written for the flow of electric current I

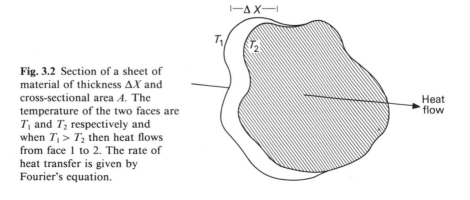

Fig. 3.2 Section of a sheet of material of thickness ΔX and cross-sectional area A. The temperature of the two faces are T_1 and T_2 respectively and when $T_1 > T_2$ then heat flows from face 1 to 2. The rate of heat transfer is given by Fourier's equation.

through a conductor, driven by a voltage difference V across the ends of the conductor (Chapter 10, p. 226).

$$I = \text{constant} \times V$$

and in this case the constant is the electrical conductance which is the reciprocal of resistance.

Heat transfer in good conductors is in fact effected by the flow of electrons and good conductors of heat also conduct electricity well. In insulators, heat flow takes place by the vibrations of the constituent molecules, a much less efficient process. Crystalline substances are good conductors of heat and electricity and this explains why ice is a better conductor than water (Table 3.2).

As an example in heat conduction, consider the heat loss from a small room at 25°C to the outside at 0°C through a window of area 2 m^2. The window thickness is 4 mm. Assuming all the heat loss takes place via the window it is given by

$$H = KA\,\Delta T/\Delta X$$
$$= 1.0 \times 2 \times 25/4 \times 10^{-3}$$
$$= 12.5 \times 10^3 \text{ or } 12.5 \text{ kW.}$$

From this prohibitively massive heat loss it is obvious that some vital factor has been omitted from the calculations. This factor is the so-called stagnant or *unstirred layers* of air that surround every body. It is very difficult to estimate the magnitude of these, but suppose they are only 2 mm thick on either side of the window, the heat loss will then be cut to approximately 0.2 kW. (Compare K for air and glass from Table 3.2) The above calculation shows the importance of the trapped air space in double-glazed windows.

When the stagnant layers are disturbed, e.g. by a wind blowing on the external surface of the window, the heat loss is much greater and in fact takes place by *forced convection*.

You can have a personal experience of the insulating power of an un-stirred layer in a sauna bath where the air temperature is over 100°C, far above that required to produced severe burns. If you blow on your arm and disturb the layers then you will feel a burning sensation. These layers will also help retain heat when you emerge to the snow outside. If,

Substance	K
Air (at STP)	0.024
Water	0.59
Brick	0.6
Glass	1.0
Ice	2.1
Silver	418

Table 3.2. Thermal conductivity K. The units of K are W m^{-1} deg^{-1}.

however, the full therapeutic value of a sudden temperature shock is required, they can be disturbed by a suitable application of birch twigs.

3.7 Heat transfer by convection

Convection is a means of heat transfer in fluids and takes place through a movement of the fluid itself. When heat energy is applied at a point in a fluid, the kinetic energy of the molecules in that region will increase, reducing the density of the fluid. The less dense region will rise, to be replaced by colder, denser fluid and in this way convection currents are set up.

The rate of transfer of heat energy from a body immersed in a fluid depends on the surface temperature of the body, its shape, surface characteristics and size and on the fluid temperature and its rate of flow relative to the body.

Convection currents surrounding objects of biological interest can be visualized using Schlieren optics, which permit small differences in refractive index to be detected. In the example shown in Fig. 3.3 the warmed air is seen to be rising from hot regions in the body. This type of exchange is termed *natural convection*. When the fluid is forced past the body, or from the body in the case of the rabbit's nostrils, then *forced convection* occurs. The pattern of air flow around a body may be either *laminar* at lower velocities, or turbulent at high velocities for both

Fig. 3.3 Schlieren photographs of a rabbit's head showing regions of relatively warm air (light) and cooler air (dark). Note the disturbed air round the nostrils, the separation of rising air over the eyebrows, and the evidence of strong heating of air round the ears (from Monteith, 1973)

forms of convection. Over a standing man, for example, air flow due to natural convection is rapid and turbulent in the region of the head, but laminar around the lower limbs. The rate of heat loss by natural convection from a naked standing man whose skin temperature is 10°C above air temperature may be 30–40 W m^{-2}, which is a large fraction of the resting metabolism.

The important factor determining the heat loss from a body immersed in a fluid is in fact the thickness (δ) of the unstirred layer of fluid surrounding the body (see above). The heat transfer per unit area (C) is given by:

$$C = K (T_s - T)/\delta$$

where C has units W m^{-2}, K is the thermal conductivity of the fluid, T_s is the surface temperature and T is the fluid temperature. δ is determined by the size and geometry of the surface and the way in which fluid circulates over it.

3.8   Forced convection

When a pump forces water round a heating system, or a fan blows on the surface of a hot body, then the heat transfer is by forced convection. Newton's law of cooling describes the rate of heat loss C from a body of unit surface area in a cool air stream, for example

$$C = K_c (T_1 - T_2) \tag{3.5}$$

when T_1 is the temperature of the body, T_2 the air temperature and K_c is the convection coefficient, with units W m^{-2} deg^{-1}.

Let us consider for a moment the processes involved in heat transfer by forced convection, e.g. at the surface of a leaf. The transfer will depend on whether or not the flow is laminar (Chapter 5, p. 79) across the leaf surface; hence from equation (5.16) it will depend on the density, viscosity, and velocity of the air. Heat will be removed from the surface of the leaf by conduction and it is then removed from the vicinity by convection. K_c therefore depends on how far laminar flow extends outwards from the surface and it depends on the heat capacity and thermal conductivity of the cooling fluid stream. For a flat plate in air (Gates, 1965),

$$K_c = 4 \sqrt{\frac{v}{L}} \text{ W m}^{-2} \text{ deg}^{-1 *} \tag{3.6}$$

where v is the stream velocity and L is the length dimension of the plate.

---

*This equation is dimensionally inhomogeneous and in fact the 4 here has units (Gates, 1965).

This means that the smaller the surface dimension along the direction of flow the greater will be the heat loss, because heat transfer takes place mainly at the edge of the plate. Many leaves are divided into lobes and this increases the rate of convective heat transfer. It will be shown in Problem 3.5 that forced convection is probably the most efficient way of removing excess heat from a leaf surface.

The convection coefficient for a cylinder in air is given by

$$K_c = 9(\mathbf{v}/D^2)^{1/3} \text{ W m}^{-2} \text{ deg}^{-1} \qquad (3.7)$$

where $\mathbf{v}$ is the air velocity and D the diameter of the cylinder.

Calculations based on Newton's Law of Cooling indicate that forced convection is also the most efficient way of removing excessive heat from insects in flight.

3.9   Heat transfer by radiation

Two bodies at different temperatures will exchange heat even when there is no possibility of exchange by conduction or convection (e.g. *in vacuo*); the transfer of heat takes place in fact by *radiation*. Good radiators of heat are also good absorbers and the most efficient absorber possible, called a *black body*, is one which will absorb all the radiation falling on it. Such a body also reradiates all the energy falling on it. The human skin, white or black, is a good approximation of a black body.

The total emissive power of a body, ϵ_0, is defined as the total radiant energy of all wavelengths emitted by the body per square metre of its surface per second.

For a black body the total emissive power is proportional to the fourth power of the temperature (Kelvin), i.e.

$$\epsilon_0 = \sigma T^4. \qquad (3.8a)$$

This is the *Stefan–Boltzmann Law* and the universal constant σ, called Stefan's constant, has the value 5.67×10^{-8} W m^{-2} K^{-4}. The total emissive power of any other body ϵ is a fraction of this and this fraction is called the *emissivity* (e) of the body, i.e.

$$\epsilon = e\,\epsilon_0. \qquad (3.8b)$$

If a beam of radiant energy from a black body is dispered into a spectrum, and if the energy content at any wavelength is measured, a maximum in energy is found and this maximum is dependent on the temperature of the radiating body (Fig. 3.4).

The energy emitted from unit area of the emitting surface *in unit time* within a small wavelength range dλ, centred on the wavelength λ, is E_λ dλ (Sears & Zemansky, 1964) where

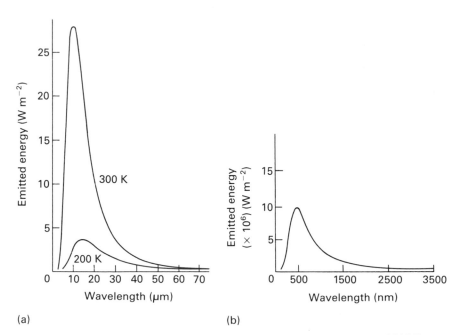

Fig. 3.4 Black-body emission curves or (a) two bodies with temperatures of 300 K and 200 K, and (b) the sun at 6000 K. Note the contrasting scales of emitted energy in units of W m^{-2} and the contrasting wavelength scales between (a) and (b). (From Barry & Chorley, 1986.)

$$E_\lambda \, d\lambda = \frac{C_1}{\lambda^5 \left(e^{C_2/\lambda T} - 1\right)} . d\lambda \text{ (Planck's formula)} \qquad (3.9)$$

$E_\lambda \, d\lambda$ has units W m^{-2}, T is the absolute temperature, and C_2 is the constant 1.44×10^{-2} m K. C_1 is another constant whose value is 3.74×10^{-16} W m^2.

The value of λ at which the energy emitted is a maximum is approximately given by

$$\lambda_{max} = \frac{0.288}{T} \times 10^{-2} \text{ m.} \qquad (3.10)$$

This shift in relative intensities accounts for the change in colour of a body emitting visible light as its temperature is raised. At 2000 K a body is red hot and at 6000 K, a body is white hot. Above 10 000 K blue light is emitted with a greater intensity than red and a body is then blue hot (some stars show this effect).

An important example of radiative exchange between two bodies occurs in the case of a baby exchanging heat energy with the walls of its incubator. Most hospital incubators are made of perspex which is in fact opaque to long wavelength (infra-red) radiation. The walls will therefore absorb energy radiated from the baby and re-radiate this to the

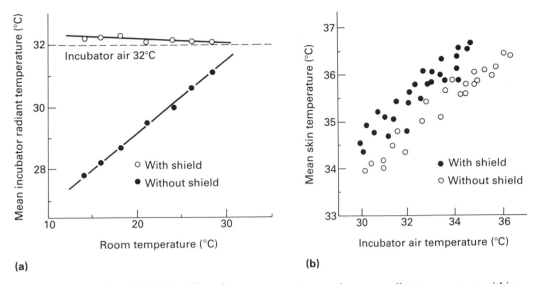

Fig. 3.5 (a) The effect of room temperature on the mean radiant temperature within an incubator, with and without an internal shield at the temperature of the incubator air. (b) The effect of a radiant shield on the mean skin temperature of 13 healthy human infants in an incubator, over a range of incubator air temperatures. Room temperature 20–22°C (after Mount, 1979).

surrounding room. The wall temperature is normally mid-way between the temperatures of the incubator air and room air, so that in a cool room the baby's radiant heat loss can be considerable. A thin perspex shield placed between the baby and the wall reduces this loss since the shield tends towards the temperature of the incubator air and hence it is a much warmer radiant environment than the incubator wall (Fig. 3.5a). Use of such a shield maintains the baby's surface temperature (Fig. 3.5b) and significantly lowers the oxygen consumption rate from 9.0 to 7.6 ml kg^{-1} min^{-1}.

3.10 Physiological mechanisms of temperature regulation in animals

Animals show both physiological and behavioural responses when they are placed in excessively hot or cold environments. The hypothalamus is the centre for the physiological control of body temperature and it receives nerve impulses from temperature receptors situated in the skin (surface temperature) and in the hypothalamus (core temperature) itself. When the animal encounters excessive heat, thermal energy is dissipated by sweating or panting and heat conservation is decreased by peripheral vasodilation. The animal also actively seeks shade.

When the core or surface temperature falls, then heat is conserved by pilo erection, which traps an insulating layer of air in the coat of the animal, and by peripheral vasoconstriction. The animal also actively increases heat production by shivering or by metabolizing brown fat (non-shivering thermogenesis, or NST). This latter process has been known for some time to play an important role in temperature regulation in hibernating animals and new-born infants (who do not show a shivering reaction in response to cold) but it is only relatively recently that its importance in response to a quite different physiological insult has been demonstrated.

3.11   Role of brown fat metabolism in the control of obesity

Normally animals control their food intake precisely, but they can be made to overeat (man provides a good example of this) by being offered a range of palatable food items (cafeteria feeding). In a series of elegant experiments, Rothwell & Stock (1979) showed that while cafeteria-fed rats consumed 80% more energy than their control-fed counterparts, their weight gain was only 27% greater than the controls. Ninety-three per cent of their additional weight gain in fact consisted of fat deposition (Table 3.3) and it is the metabolism of this fat (measured by the animal's oxygen consumption) which not only raises the animal's temperature, but also reduces obesity. Interestingly, this metabolism appears to be under the control of the sympathetic nervous system as it can be stimulated by adrenaline and inhibited by the adrenergic blocking agent propranolol (Table 3.4).

Brown adipose tissue is laid down in discrete areas in rat and in man and the highest density is in the neck region and near the scapulae. When the skin temperature is measured in man, by infra-red camera

	Control	Cafeteria
Initial body weight (kg)	0.318	0.315
Body weight gain (kg)	0.103	0.131
Body fat gain (kg)	0.040	0.066
Energy intake (kJ)	6480	11 670
Final body energy (kJ)	1790	2230
Energy expenditure (kJ)	4690	9440

Table 3.3 This table gives the energy balance data for control and cafeteria-fed rats over a 21-day period. The energy intake was estimated from the weight of the stock and cafeteria diets and their metabolizable energy densities. Several animals were killed on day 21 and their body composition determined by analysis. The body energy was calculated using values of 38.0 and 19.6 MJ kg^{-1} for the energy value of fat and protein respectively.

A group of six animals were also killed on day 0, the body energy content determined in the same way and this value subtracted from the final body energy value of the experimental animals in order to estimate body energy gain. Energy expenditure over the 21 days was estimated by subtracting the body energy gain from the metabolizable energy intake.

	Control	Cafeteria
(a) Resting oxygen consumption	10.9	14.1
After noradrenaline injection	16.5	28.2
After propranolol injection	10.9	11.8
(b) Rectal temperatures (°C)		
Resting temperature	37.5	38.0
After noradrenaline	37.8	39.3
(c) Scapular temperature (°C)		
Resting temperature	36.0	36.8
After noradrenaline	36.2	38.2

Table 3.4. (a) Oxygen consumption (ml min^{-1} kg^{-1} body weight) (b) rectal, and (c) scapular temperature data from control and cafeteria-fed rats showing the importance of sympathetic nervous system control of brown fat metabolism

techniques (Fig. 3.6) it is precisely these regions which 'light up' when metabolism is stimulated by the sympathomimetric agent ephedrine.

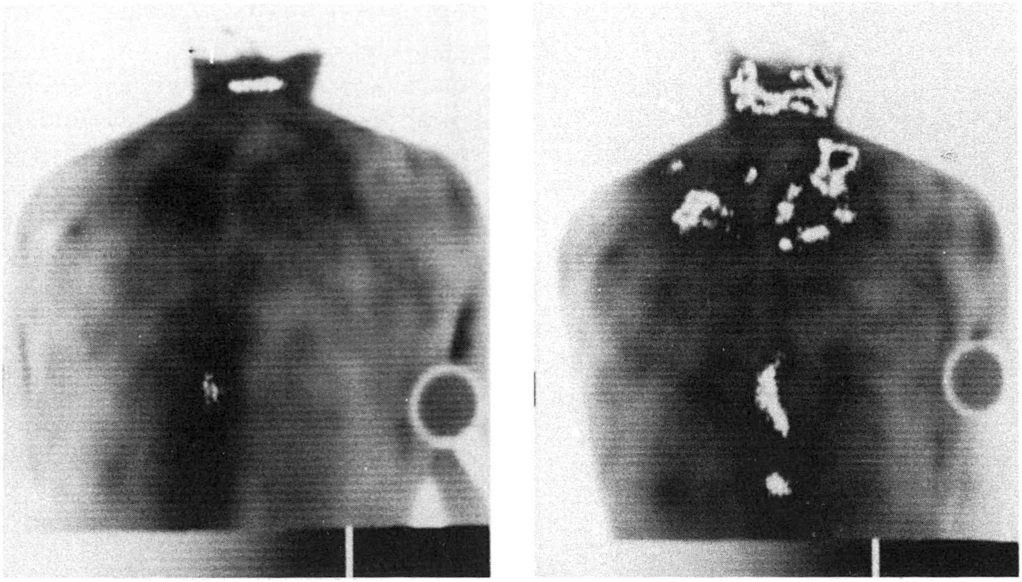

(a) (b)

Fig. 3.6 Infra-red thermograms of a male subject (age 36 years) before and after administration of ephedrine. Before taking ephedrine (a) the highest skin temperature recorded was 34.1°C. The areas of skin isothermal to this temperature appear as white patches on the image. (b) One hour after taking ephedrine the areas at this temperature have increased and are concentrated between and above the scapulae and in the neck region. (Photographs kindly provided by Professor M.J. Stock and used with his permission.)

Problem 3.1

(a) A lizard of mass 3×10^{-3} kg in bright sunlight casts a shadow of 1.9×10^{-4} m^2 on a piece of paper held perpendicularly to the sun's rays. If the incident energy from the sun was 1×10^3 $J\,m^{-2}\,s^{-1}$, measured at the paper, calculate the maximum possible rate of rise of the lizard's temperature (specific heat of lizard = 4.2×10^3 J kg^{-1}°C^{-1}).

(b) How does the lizard in fact maintain a stable temperature?

Problem 3.2

A mountain climber has a body surface area of 1.8 m^2 and wears fibrous clothing 10^{-2} m thick. He has a skin temperature of 33°C while the outer surface of his clothing is at 0°C. Calculate the total rate of conduction of heat through his clothing in watts, (a) taking the thermal conductivity of his dry clothing to be 0.04 W m^{-1} K^{-1}, (b) assuming that the clothing is wet through and that the appropriate thermal conductivity is that of water (0.6 W m^{-1} K^{-1}).

Problem 3.3

The beetle, *Melanophila acuminata*, lays its eggs in conifers freshly killed by forest fires. Its metathorax has receptors that are especially sensitive to infra-red radiation and it can detect and orient towards a 50 acre area of glowing wood at a distance of 6 km.

Estimate (a) the total rate of energy radiation (in watts) from such an area, assuming the wood to be glowing orange and behaving as a black body at a temperature of 1000 K; (b) the wavelength corresponding to the peak of the spectral curve at this temperature; (c) the intensity (in W m^{-2}) of the radiation incident on the beetle at the distance of 6 km. Assume for simplicity that the radiation from the wood is distributed evenly over a hemisphere with its centre at the wood and the beetle at its circumference (2.5 acres = 10^4 m^2 and Stefan's constant = 5.7×10^{-8} W m^{-2} K^{-4}).

Problem 3.4

A recent report on food preservation in Britain revealed that the temperature of meat in transparent plastic packs could be as much as 12°C above the freezer temperature when they were stored in a situation where they were exposed to illumination. (a) Explain how this high temperature effect occurs. (b) Design a pack which would prevent this.

Problem 3.5

The total energy falling on a horizontal leaf from the sun and the immediate environment is 800 W m^{-2}.

(a) If we asume that no heat is lost or reflected from the leaf, what would be the rate of rise of the leaf's temperature, given that the specific heat and weight per unit area of the leaf are $3.8 \times 10^3 \, J \, kg^{-1} \, K^{-1}$ and $0.1 \, kg \, m^{-2}$?

(b)   Think for a moment about the possible ways in which the leaf might get rid of this unwanted energy.

(c)   Calculate the temperature the leaf would reach if it lost all its heat by radiation.

(d)   As you will now have shown that the temperature of the leaf is still too high, calculate the transpiration rate necessary to maintain the temperature at the reasonable level of 40°C.

(e)   If a wind velocity of $4.5 \, m \, s^{-1}$ (10 mph) is blowing across the leaf, calculate the fraction of the total incident radiation that would be lost by forced convection when the temperature difference between the leaf and its environment is (i) 5°C (ii) 10°C. Assume the leaf to be a square plate of linear dimension 0.01 m.

Answer

(a)   Substitute the values given above in equation (3.3), hence

$$\frac{\Delta T}{\Delta t} = \frac{800}{0.1 \times 3.8 \times 10^3} \, deg \, s^{-1}$$
$$= 2.1 \, deg \, s^{-1}$$

which is a very fast temperature rise.

(c)   If we assume that heat is lost by radiation alone, then thermal equilibrium will be reached when this balances the incoming radiation, i.e. when

$$\sigma T^4 = 800 \, W \, m^{-2}$$
$$or \; T = \sqrt[4]{\frac{800}{5.7 \times 10^{-8}}} \, K$$
$$= 344 \, K \; or \; 71°C.$$

This temperature is greater than the highest recorded from a leaf (49°C) and so there must be some additional cooling process.

(d)   Remember that the leaf will be reradiating some of the incoming energy, so the energy which has to be lost by transpiration will be the difference between the two.

Loss due to reradiation at 40°C (313 K) $= 5.7 \times 10^{-8} \times (313)^4 \, W \, m^{-2}$
$$= 550 \, W \, m^{-2}.$$

Hence the amount to be lost by transpiration
$$= 250 \, W \, m^{-2}.$$

Since 2.5×10^6 J are required to vaporize 1 kg of water at 40°C, the excess energy will vaporize 10^{-4} kg m^{-2} s^{-1}. This is the required transpiration rate and it is a relatively high one. There is, however, another possible method for heat loss.

(e) The heat loss carried by forced convection is given by

$$\Delta Q = K_c \Delta T \text{ (equation 3.6)}$$
$$= 424 \text{ W m}^{-2} \text{ when } \Delta T = 5°C$$
$$\text{and} \quad = 848 \text{ W m}^{-2} \text{ when } \Delta T = 10°C.$$

You should now have sufficient insight to investigate temperature regulation in other systems. An interesting starting point is the question of heat loss and body temperature in flying insects studied by Church (1960).

References

Barry, R.G. & Chorley, R.J. (1968) *Atmosphere, Weather and Climate.* Methuen, London.

Church, N.S. (1960) Heat loss and the body temperatures of flying insects. *Journal of Experimental Biology*, **37**, 171–212.

Gates, D.M. (1965) *Energy Exchange in the Biosphere.* Harper & Row, London.

Monteith, J.L. (1973) *Principle of Environmental Physics.* Arnold, London.

Rothwell, N.J. & Stock, M.J. (1979) A role for brown adipose tissue in diet-induced thermogenesis. *Nature*, **218**, 31–35.

Sears, F.W. & Zemansky, M.W. (1964) *University Physics.* Addison–Wesley, Reading, Mass.

Further reading

Burns, D.M. & MacDonald S.G.C. (1970) *Physics for Biology and Pre-Medical Students.* Addison–Wesley, London.

Jarman, M. (1970) *Examples in Quantitative Zoology.* Arnold, London.

Mount, L.E. (1979) *Adaptation to Thermal Environment.* Arnold, London.

Phillipson, J. (1966) *Ecological Energetics.* Arnold, London.

Singleton, W.T. (1982) *The Body at Work.* Cambridge University Press, England.

4 Fluids: Pressure and Gases

4.1 Introduction

We live in a fluid environment; the air flows round us, conveying the all-important oxygen to our blood, which in turn carries it to our tissues. Even the seemingly solid earth beneath us is in a constant state of flux. All *flows* of materials are set up by *driving forces* and we shall spend a large fraction of the following two chapters investigating the relationships between the forces and their induced flows. First, however, we shall discuss the static properties of fluids and the relationships between pressures and volumes.

4.2 Pressure

A fluid in a container exerts a force on every part of the container it is in contact with, because every molecule of the fluid is in a continual state of motion. When a fluid molecule bumps against the container walls, its velocity is altered, and so it must exert some force. The fluid in fact exerts a *pressure*.

The pressure at any point is defined as the force per unit area surrounding the point:

$$\text{pressure} = \text{force/area} \tag{4.1}$$

and has units $N\ m^{-2}$*.

One common way of measuring air pressure is by means of the mercury barometer. When a glass tube is filled with mercury and inverted over a mercury bath (Fig. 4.1a), the mercury level falls until the column is in equilibrium with the air pressure. Considering an imaginary cut through B, parallel to the surface of the bath, we can then see that as mercury is incompressible, the pressure exerted on the upper surface of the cut by the lower is the atmospheric pressure, P. If the cross-sectional area of the cut is A, then this upward force will be PA. The downward force due to the mass of mercury above the cut is $m\mathbf{g}$ and at equilibrium

$$
\begin{aligned}
PA &= m\mathbf{g} \\
&= \rho A\mathbf{g}h
\end{aligned}
\tag{4.2}
$$

---

*The unit of pressure in the SI system is the Pascal (Pa).

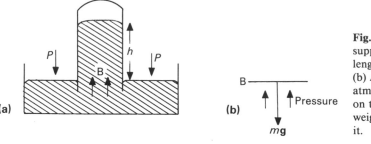

Fig. 4.1 (a) Air pressure can support a column of mercury of length h in an enclosed tube. (b) At the surface through B the atmospheric pressure P acting on the area A balances the weight of the column above it.

where ρ is the density of the mercury and h the height of the column. Hence

$$P = \rho gh. \tag{4.3}$$

At sea-level, the average height which the air pressure can support is 0.76 m of mercury, and as the density of mercury is 13.6×10^3 kg m^{-3}, the atmospheric pressure is approximately 10^5 N m^{-2}. In older texts, pressures are often given in terms of atmospheres or metres of mercury.

Problem 4.1

The mean arterial pressure at the heart level in man is often quoted as 100 mmHg.
(a) Convert this into SI units.
(b) What is the pressure (i) at a point 0.5 m above the heart and (ii) at the feet, some 1.5 m below the heart (density of blood can be taken as 10^3 kg m^{-3}).

4.3   Ideal gases

The pressure P exerted by a gas depends on the volume V and temperature T of the gas, and on the number n of moles present. The relationship between the quantities can be measured experimentally. If the temperature is fixed during the experiment, and if the relationship PV/nT is plotted against pressure, then a smooth curve T_1, is obtained (Fig. 4.2). When the process is repeated at other temperatures T_2 and T_3 it is found that all the *isotherms* intersect the vertical axis at the same point.

It is found that this limiting value ($P=0$) is the same for all gases. The point of intersection is called the *Universal Gas Constant R* and has the value

$$R = 8.3 \text{ J mol}^{-1} \text{ deg}^{-1}.$$

It is customary to define an ideal gas as one for which PV/nT is equal to R at all pressures. The relationship

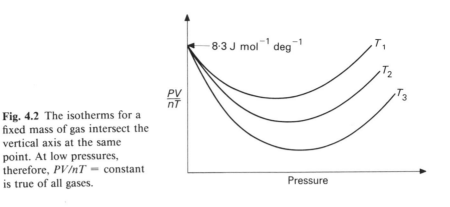

Fig. 4.2 The isotherms for a fixed mass of gas intersect the vertical axis at the same point. At low pressures, therefore, PV/nT = constant is true of all gases.

$$PV/nT = R \qquad (4.4)$$

is also called the equation of state for an ideal gas.

If we have a mixture of ideal gases, the total number of moles n is given by

$$n = n_1 + n_2 + n_3 \qquad (4.5)$$

and as $P = \dfrac{nRT}{V}$

$$P = \frac{n_1 RT}{V} + \frac{n_2 RT}{V} + \frac{n_3 RT}{V}$$

$$= P_1 + P_2 + P_3 \qquad (4.6)$$

$$= \sum_i P_i.$$

Where P_i, called the *partial pressure*, is the pressure which the gas i would exert were it alone in the volume V. Equation (4.6) is called Dalton's Law of Partial Pressures. From the ideal gas equation, for a fixed mass of gas, nR is constant and so PV/T is constant, i.e.

$$\frac{P_1 V_1}{T_1} = \frac{P_2 V_2}{T_2} \qquad (4.7)$$

If the temperatures T_1 and T_2 are also the same,

$$PV = \text{constant.} \qquad (4.8)$$

This relationship was discovered experimentally by Boyle in 1660.

4.4   Properties of real gases

Boyle's Law (equation 4.8), predicts that a graph of pressure against volume should be a rectangular hyperbola, and there should be one smooth curve for each temperature. The isotherms of a real gas are

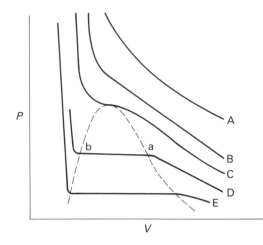

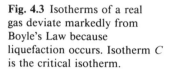

Fig. 4.3 Isotherms of a real gas deviate markedly from Boyle's Law because liquefaction occurs. Isotherm *C* is the critical isotherm.

shown in Fig. 4.3. The upper curve A corresponds to the ideal gas behaviour and the substance remains a gas at all volumes. Curves B and C show a marked deviation from the ideal, but the substance still remains a gas.

For curve D, however, the situation is different. If originally the gas is under only a low pressure and this is then gradually increased, while the temperature is held constant, then a point a is reached when the volume decreases without any further increase to pressure. Due to intermolecular interactions the gas begins to liquefy until finally a point b is reached when liquefaction is complete. Thereafter the isotherm rises steeply as liquids are virtually incompressible. Isotherm C is called the critical isotherm.

The dotted curve encloses the region where the gas exists in equilibrium with its liquid. A gas above its critical temperature is referred to as *vapour*, and the pressure corresponding to the line ab is the *saturated vapour pressure* at that temperature. The saturated vapour pressure is the maximum pressure that can be exerted by the vapour at that temperature and is dependent only on the temperature (Table 4.1).

Temperature (°C)	Vapour pressure (kN m^{-2})
0	0.60
5	0.87
10	1.21
15	1.69
20	2.30
25	3.14
30	4.18
35	5.55
40	7.26

Table 4.1. Vapour pressure of water at various temperatures.

The partial pressure of water vapour in the atmosphere at any temperature is usually less than the saturated vapour pressure for the same temperature and the ratio of the two expressed as a percentage is known as the *relative humidity*, i.e.

$$\text{relative humidity} = 100 \times \frac{\text{partial pressure of vapour}}{\text{saturated vapour pressure at same temperature}}$$

$$= 100 \times \frac{\text{amount of vapour the air contains}}{\text{amount it would contain if saturated}}$$

Saturation can be achieved in two ways.

1   The total water content can be increased until the pressure of the vapour is the saturated vapour pressure.

2   The temperature can be lowered until the actual amount of water vapour in the air is enough to cause saturation at the new temperature. It is this process that causes mist, fog, etc. The temperature at which moist air would be saturated is called the *dew point* and this constitutes an easily measured parameter. All that is necessary is to cool a brightly polished metal surface and observe the temperature at which it becomes clouded with moisture. The relative humidity is then simply the saturated vapour pressure at the dew point divided by the saturated vapour pressure at the actual air temperature times 100. For example, suppose the dew point measured in this way is 10°C when the air temperature is 20°C. We then know that the vapour in the air is saturated at 10°C, hence its partial pressure is 1.2 kN m^{-2} equal to the saturated vapour pressure at 10°C. As the pressure necessary for saturation at 20°C is 2.3 kN m^{-2},

$$\text{the relative humidity} = \frac{1.2}{2.3} \times 100 = 52\%.$$

4.5   Swim bladder

The ideal gas laws can be invoked to explain the mechanism of the swim bladder, which is a buoyancy device that enables Teleost fish to hover in mid-water with a minimum expenditure of energy. The swim bladder is a gas-filled organ within the body cavity of fish which reduces the overall density of the fish from about 1.07 × 10^3 kg m^{-3} to near 1.0 × 10^3 kg m^{-3}, the density of the medium (*neutral buoyancy*).

To a certain extent the density of a fish is under its own control as some fish can secrete gas in and out of the swim bladder, while others have a special sphincter mechanism which allows for a rapid discharge of excess gas through the mouth.

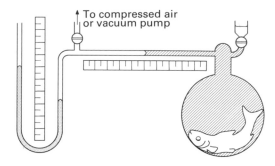

Fig. 4.4 Apparatus for investigating the mechanical properties of the swim bladders of fish. (From Alexander, 1968. Reproduced by permission of Sidgwick & Jackson Ltd.)

For any fixed mass of gas, however, there will only be one depth at which the fish will be in equilibrium, and even this equilibrium is unstable as a slight upward movement will reduce the pressure on the gas, allow it to expand and so reduce the overall density of the fish, forcing it up still further. Hence a continuous adjustment of the quantity of gas in the swim bladder may be necessary if neutral buoyancy is to be maintained. Fig. 4.4 shows the apparatus used by Alexander to study the pressure–volume relationships of the swim bladder. An anaesthetized fish is placed in a flask which is completely filled with weak anaesthetic solution and the flask has a side-arm capillary tube which is partly filled with solution. The pressure on the fish is altered by means of compressed air or a vacuum pump and the pressure is measured by means of a mercury manometer. The volume change induced by a change in pressure is calculated from the distance which the water meniscus travels along the capillary tube of known cross-sectional area.

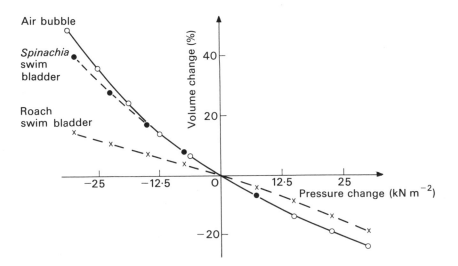

Fig. 4.5 *P–V* relationships of the swim bladder. Changes of volume of an air bubble and of the swim bladders of *Spinachia* and roach (*Rutilus*), caused by changes of external pressure. (From Alexander, 1968. Reproduced by permission of Sidgwick & Jackson Ltd.)

The results (Fig. 4.5) show that the pressure–volume relationships of the swim bladder of fish of the genus *Spinachia* follow closely the behaviour expected from the gas laws. Fish of this order in fact have swim bladders with slack, highly extensible walls. The *Cypriniformes* (of which the roach is an example) deviate from the air-bubble behaviour. This is because they have swim bladders with taut inextensible walls. This means in fact that when a roach changes its depth its density changes much less than if it had a more extensible swim bladder.

4.6 Work in changing the volume of a fluid

In mechanical systems (Chapter 2, p. 14) when a force **F** acts on the system, the work done is the product of the system's displacement dx and the component of the force $\mathbf{F}_x$ along the displacement. The sign convention of mechanics provides for positive work when $\mathbf{F}_x dx$ is positive, i.e. work done *on* the system is positive. Unfortunately, however, the thermodynamicists of old adopted the opposite sign convention because they focused their attention on the work output of the heat engines they were studying. They said that work done *by* the system is positive and although this is a convention used in many physics texts (including the admirable Sears & Zemansky, 1964) it will not be used here because it is confusing. Therefore, our sign convention is *work done on the system is positive*.

Consider a fluid contained in a cylinder equipped with a movable piston (Fig. 4.6a). Suppose the cylinder has cross-sectional area A and

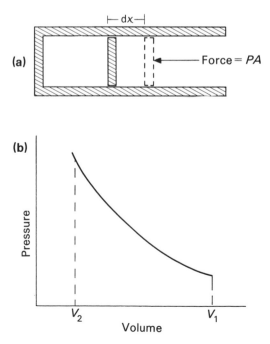

Fig. 4.6 See text for explanation.

that the pressure exerted on the system at the piston face is P. The force exerted on the system is therefore PA. If the piston moves an infinitesimal distance dx, then the work dw done by this force on the system is

$$dw = PA\,dx \tag{4.9}$$

The work done is positive as the displacement dx is in the same direction as the force but

$$A\,dx = -dV$$

where dV is the change in volume.

The negative sign occurs here as the volume decreases when dx is positive. Hence

$$dw = -P\,dV \tag{4.10}$$

and in a finite change of volume from V_1 to V_2 the total work done will be given by

$$W = -\int_{V_1}^{V_2} P\,dV \tag{4.11}$$

The total work done on the system is therefore the negative of the area under the pressure–volume curve (Fig. 4.6b).

If the pressure remains constant,

$$W = -P(V_2 - V_1) \tag{4.12a}$$

and positive work is done on the system when $V_1 > V_2$, i.e. when the fluid is compressed.

It can also be shown that when the pressure varies, but the volume of fluid remains constant, the work done on the fluid is

$$W = -(P_2 - P_1)V. \tag{4.12b}$$

4.7   Work done in an adiabatic process

Applying the First Law of Thermodynamics to an adiabatic process ($\Delta Q = 0$) we get from equation (3.2)

$$\Delta E = \Delta W. \tag{4.13}$$

Thus the change in internal energy of a system in an adiabatic process is equal to the work done. If the work done is positive, e.g. when a system is compressed, ΔW is positive, ΔE is positive and the temperature of the system increases. If ΔW is negative, e.g. when a system expands, the internal energy decreases and the temperature falls.

An adiabatic expansion is one major cause of precipitation (rain-

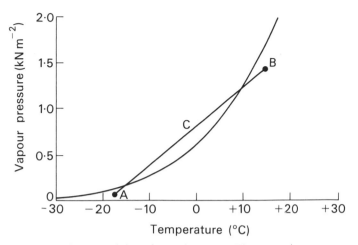

Fig. 4.7 Horizontal mixing of two air masses. The saturation vapour pressure curve as given and the horizontal mixing of two unsaturated air masses A and B to the right of the curve results in one supersaturated air mass C (after Barry & Chorley, 1968).

fall). When a parcel of air moves rapidly upwards, there is rapid expansion because of the decrease in pressure. In the rapid process there is no time for the parcel to exchange heat with the environment and so the temperature is lowered. If the temperature is lowered below the dew point, water condenses and rain falls.

The horizontal mixing of two unsaturated air masses can also produce precipitation (Fig. 4.7). Air masses to the right of the curve are unsaturated, whereas those to the left are saturated. A and B are both unsaturated, but they can combine to form a third mass C that is supersaturated at the new temperature and so will become cloudy. If there are condensation nuclei present, rain will fall.

Contact cooling is another method producing a saturated air mass. On a clear winter's night strong radiation from the land will cool the surface of the earth very quickly and this surface cooling will gradually extend to the moist lower air, reducing the temperature to a point where condensation occurs in the form of dew, fog, or frost, depending on the amount of moisture involved and the dew point value. When the latter is below 0°C it is referred to as the *hoar frost point* and the air is saturated with respect to ice.

4.8 Solubility of gases in liquids

The maximal amount of gas that can dissolve in a given liquid depends on four factors.
1   The pressure of the gas; solubility increases with pressure.
2   The temperature of the solvent; the solubility decreases with increasing temperature.

3 The solute content of the solvent; the solubility decreases with increasing solute content.

4 The nature of the solvent; the *absorption, coefficient* of a gas.

The *absorption coefficient* of a gas in a particular solvent is that volume of gas at STP that can dissolve in unit volume of the liquid at 0°C. The absorption coefficient of CO_2 in water is 1.713 atm^{-1} and this means that 1 l of water at 0°C dissolves 1.713 l of CO_2 at STP. The absorption coefficients of oxygen and nitrogen at 0°C are 0.049 and 0.024 atm^{-1} respectively.

Absorption coefficients can also be given for solvents at temperatures other than 0°C but it should be carefully noted that in these cases the amount of dissolved gas is given by the volume this dissolved gas would occupy at STP (Table 4.2).

Temperature (°C)	Absorption coefficient (oxygen in water)
0	0.0489 atm^{-1}
10	0.0380
20	0.0310
30	0.0261

Table 4.2. (*See text*)

If a gas is present at a pressure of a fraction of an atmosphere, then its solubility will be reduced by the same fraction.

For example, air at sea-level contains about 21% oxygen and although this air has a pressure of 10^5 N m^{-2}, the amount of oxygen that dissolves corresponds to a pressure of 2.1×10^4 N m^{-2}.

We can calculate the solubility of any gas if we know the absorption coefficient a, the partial pressure p, and the pressure P of the total gas mixture.

Note that it is now usual to quote pressures in the SI units, hence when

$$P = 1 \text{ atmosphere } (10^5 \text{ N m}^{-2})$$

$$V = \frac{ap}{10^5} \text{ litres per litre} \tag{4.14}$$

and V in this case is the volume in litres of the gas reduced to conditions of STP, that dissolves in 1 litre of solvent.

Problem 4.2

Lake Titicaca lies at 4000 m above sea-level between Peru and Bolivia and the barometric pressure is about 59.2 kN m^{-2}. How much oxygen can dissolve in 1 litre at 20°C? (after Jarman, 1970).

Answer

(a) $pO_2 = \dfrac{21}{100} \times (59.2 - 2.3) = 12\,\text{kN m}^{-2}$

$\therefore V = \dfrac{0.031 \times 12 \times 10^3}{10^5}$

$= 3.6 \times 10^{-3}\,\text{litres.}$

(b) Similarly, show that there is about $6.4 \times 10^{-3}\,\text{l}$ of oxygen per litre of Norfolk Broad water (at sea-level), making the somewhat doubtful assumption that the water contains no salt or dissolved organic material.

Note: atmospheric pressures must be corrected for saturated water vapour pressure.

4.9 Osmotic pressure

The concept of *osmotic pressure* is a very widely used one and in order to understand it we have to go to the somewhat artificial system shown in Fig. 4.8.

A membrane M separates two sucrose solutions, the solution on the right side being more concentrated than that on the left. Initially, the levels in the two arms are at the same height, but if the membrane is what is called a *semipermeable* one, i.e. it allows the passage of water

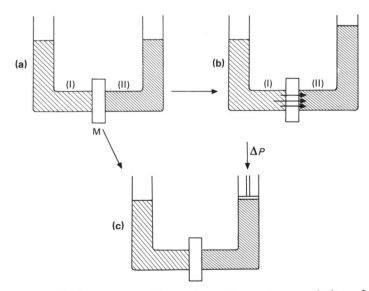

Fig. 4.8 (a) An ideally semipermeable membrane M separates two solutions of different solute concentrations ($C_{II} > C_I$). (b) Experimentally we find that water moves from phase I to II. (c) This net movement can be prevented by applying a pressure ΔP on phase II.

only and not solute, then the level in the right tube is found to rise. This means that there is a flow of water from left to right, and the driving force of this flow is the difference in the *water potential* between the two sides. The water potential of a phase depends on two factors: (i) the concentration of solute dissolved in the phase, and (ii) the hydrostatic pressure on the phase. The presence of dissolved species lowers the water potential, whereas application of a pressure raises the potential.

$$J_W = L_p (P_I - P_{II}) - k (C_I - C_{II}) \qquad (4.15)$$
$$= L_p \Delta P - k \Delta C \qquad (4.16)$$

where J_W is the flow of water; P_I and P_{II} are the pressures on the semipermeable membrane in phases I and II respectively; C_I and C_{II} are the concentrations of solute in phases I and II respectively; and k and L_p are two constants. The negative sign arises because water flows from a phase where the concentration of dissolved solute is low, to one where it is high.

Now, it can be shown, both from theory and from experiment, that for an ideally semipermeable membrane and ideal solutions

$$k = RTL_p$$

Hence equation (4.16) can be written in the form

$$J_W = L_p (\Delta P - RT \Delta C) \qquad (4.17)$$
$$= L_p (\Delta P - \Delta \pi) \qquad (4.18)$$

where $\Delta \pi = RT \Delta C$ is called the osmotic driving force or osmotic pressure and it is determined by the sum of *all* the solute molecules and individual ions in the phase; $(P_I - \pi_I)$ and $(P_{II} - \pi_{II})$ are the water potentials (ψ) of phases I and II respectively. L_p is often referred to as the osmotic permeability.

When the pressure difference between the two phases is adjusted so that there is no net water flow, i.e. $J_W = 0$, then

$$\Delta P = \Delta \pi \qquad (4.19)$$

and the pressure required to balance the osmotic flow of water is called the *osmotic pressure*.

The *osmolality* of a solution is the concentration of dissolved, dissociated solute, e.g. a 155 mM solution of NaCl is said to contain 310 milliosmoles per litre or strictly speaking 310×10^3 milliosmoles per cubic metre. One solution is said to be hyperosmotic, isosmotic or hyposomotic with respect to another, when it contains a higher, the same, or a lower concentration of dissolved solute respectively. A 155 mM solution of NaCl is approximately isosmotic with respect to blood. Raisins swell when soaked in tap water because they are hyperosmotic with respect to the water.

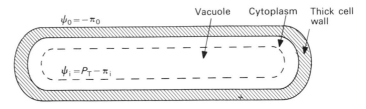

Fig. 4.9 The thick cellulose wall enables the relatively concentrated cytoplasm to remain in osmotic equilibrium with the dilute aqueous environment. (After Thain, 1967.)

4.10   Osmotic phenomena in plants

While most animal cells are isosmotic with the blood that bathes them, many plant cells are in a dilute aqueous environment, pond water for example, where the osmotic pressure is much lower than in the plant cell contents (vacuole and cytoplasm). Osmotic equilibrium can be achieved between the two phases, however, as the thick cellulose plant cell wall exerts a *turgor pressure* (P_T) (Fig. 4.9).

At equilibrium the water potentials of the internal and external phases are equal, i.e.

$$\psi_i = \psi_o \tag{4.20}$$

or

$$P_T - \pi_i = -\pi_o. \tag{4.21}$$

The water potentials of plant cells, or indeed any cell, can be measured by immersing them in a series of solutions of a non-penetrating solute such as sucrose. From equation (4.19) the water potentials of the cells are equal to the osmotic pressures of the solutions in which the cells neither gain nor lose water. Uptake or loss of water may be detected by changes in weights or volumes of the cells.

See also the possible part played by osmosis in phloem translocation (Chapter 5, p. 83).

Problem 4.3

A certain tribe of South American Indians use a blowpipe that is exactly 2 m long and has a smooth bore with a diameter of 1 cm. The dart has a mass of 5 g and a piston end that just fits the bore. If the Indians can maintain an expiratory pressure of 10 kPa, calculate (1) the acceleration of the dart, and (2) the velocity as it leaves the blowpipe when aimed in a horizontal direction.

Many tribes use blowpipes that seem inordinately long relative to the size of their owners. Is there any advantage in having a long blowpipe (apart from dislodging fruit from trees)?

References

Alexander R.M. (1968) *Animal Mechanics*. Sidgwick & Jackson, London.
Barry R.G. & Chorley R.J. (1968) *Atmosphere, Weather and Climate*, Methuen, London.
Jarman M. (1970) *Examples in Quantitative Zoology*, Arnold, London.
Sears F.W. & Zemansky M.W. (1964) *University Physics*. Addison–Wesley, Reading, Mass.
Thain J.F. (1967) *Principles of Osmotic Phenomena*. Monographs for Teachers No. 13. Royal Institute of Chemistry.

Further reading

Alexander R.M. (1971) *Size and Shape*. Arnold, London.
Burns D.M. & MacDonald S.G.G. (1970) *Physics for Biology and Pre-Medical Students*. Addison–Wesley, London.
Dainty, J. (1963), Water relations of plant cells. *Advances in Botanical Research*, **1**, 279–326.
Dick D.A.T. (1966) *Cell Water*, Butterworth, London.
House C.R. (1974) *Water Transport in Cells and Tissues*. Arnold, London.

5 Fluid Flow and Viscosity

5.1 Introduction

Fluids flow as the result of driving forces, e.g. in the case of a pipe, flow occurs when there is a pressure difference across the ends of the pipe. This simple basis is the starting point for investigation into systems as widely diverging as blood flow through veins and arteries and phloem transport in plants. In this discussion we shall be concerned initially with *ideal fluids* which are considered to be *incompressible*.

5.2 The continuity equation

Consider a fluid flowing along a tube whose cross-sectional area varies from point to point (Fig. 5.1). At position (1) the cross-sectional area is A_1 and the linear velocity is v_1, while at (2) they are A_2 and v_2 respectively. As the fluid is considered to be incompressible, the density ρ will not vary.

As no fluid leaves through the walls, the mass flowing in across A_1 in a time dt must equal that leaving A_2 in the same time. Hence

$$\rho A_1 v_1 dt = \rho A_2 v_2 dt$$

and so

$$A_1 v_1 = A_2 v_2. \tag{5.1}$$

This obviously implies that as the vessel narrows, the linear velocity of the fluid through it increases.

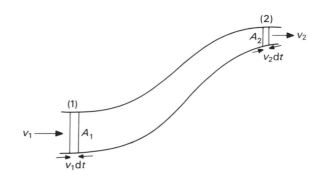

Fig. 5.1 Flow through a tube of varying cross-sectional area. At positions (1) and (2) the cross-section areas are A_1 and A_2 and the velocities v_1 and v_2 respectively.

Problem 5.1

Does the continuity equation apply to the following data obtained for the cardiovascular system in man?

	Total cross-sectional area of vessels (m²)	Linear velocity of blood (m s⁻¹)
Aorta	2.5×10^{-4}	30×10^{-2}
Capillary bed	1.9×10^{-2}	5×10^{-3}
Vena cava	1×10^{-3}	8×10^{-2}

5.3   Bernoulli's equation

When an incompressible fluid flows along a tube of varying cross-section its velocity changes. It must therefore be acted on by a resultant force, and this means in fact that pressure may vary along the tube. Let us consider the general case where the height of the tube above some reference level also changes (Fig. 5.2).

Consider an element of the fluid at positions (1) and (2). The pressures at the two positions are P_1 and P_2 respectively, y_1 and y_2 are the heights of the centres of fluid mass above the earth's surface, and $\mathbf{v}_1$ and $\mathbf{v}_2$ are the velocities of the fluid. The net work done on moving an element of volume V from position (1) to (2) is $(P_1 - P_2) V$ (equation 4.12b) and this must equal the change in potential and kinetic energies of the element, i.e.

$$(P_1 - P_2) V = mgy_2 - mgy_1 + \tfrac{1}{2}m\mathbf{v}_2^2 - \tfrac{1}{2}m\mathbf{v}_1^2$$

as $V = m/\rho$

then $P_1 - P_2 = \rho g y_2 - \rho g y_1 + \rho \mathbf{v}_2^2 - \rho \mathbf{v}_1^2$

or $P + \rho g y + \tfrac{1}{2}\rho \mathbf{v}^2 = \text{constant}$ (5.2)

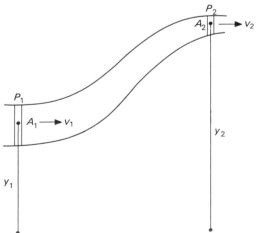

Fig. 5.2 The flow through a tube of varying cross-sectional area as in Fig. 5.1, and y_1 and y_2 are the heights of the tube above some arbitrary reference level.

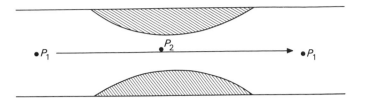

Fig. 5.3 The pressure at a point within the constricted area is less than the pressure in the normal vessel ($P_2 < P_1$) and this causes flutter, a medical condition.

In equation (5.2), $\frac{1}{2}\rho v^2$ has the units kinetic energy per unit volume. This expression is sometimes referred to as the *kinetic energy density* of the fluid and it also has the units of pressure.

In most problems of biological interest, $y_1 = y_2$ and Bernoulli's equation becomes

$$P + \tfrac{1}{2}\rho v^2 = \text{constant.} \tag{5.3}$$

As the velocity increases with decreasing area (equation 5.1), equation (5.3) implies that the pressure in a tube decreases at a point of constriction (Fig. 5.3) and this has important consequences in blood vessels.

In diseased arteries, an *atherosclerotic plaque* can narrow the lumen to under one-fifth of the original area, leading to a greatly reduced arterial pressure in this section of the vessel. If in fact the arterial pressure is lowered below the critical closing pressure, the artery will close (Fig. 5.4). However, when the flow is reduced to zero, the kinetic energy component disappears, the pressure builds up once more and the cycle is repeated. This sequence of events is called *flutter*, and the sounds produced by this vibratory motion are used diagnostically.

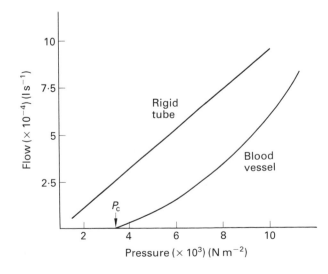

Fig. 5.4 The relationship between pressure and flow in an artery. It is not linear because of the elastic properties of the walls, and a certain pressure P_c is required to overcome the elasticity and open the vessel.

So far it has been possible to derive two very valuable equations that apply to the gross movements of fluids. However, we cannot analyse in detail the flow through capillaries for example without taking into account the fluid viscosity.

5.4   Pressure and flow in tubes

In order to drive a constant flow of liquid between two positions, for example along a river or along a pipe, a constant pressure difference must be maintained in order to overcome the internal friction or *viscosity* of the fluid. Newton was the first to ascribe the viscosity to a lack of slipperiness as the different parts of the fluid passed one another. This hypothesis lead to the concept that the movement of fluid is effected by infinitesimally thin laminae sliding over one another.

The regularity of fluid movement can be seen if a series of straws are dropped in a line across a smooth flowing river, and their positions are noted at successive times (Fig. 5.5a). It is found that the straws in the middle move furthest while those nearest the bank scarcely move at all. The straws map out the velocity profile of the flow at the surface and it consists of *streamlines*, i.e. lines joining points of constant velocity, running parallel to the banks. Adjacent streamlines have different velocities and so a force has to be exerted to allow the lines to slip over one another. Regions of constant velocity extend into the bulk of the fluid making up laminae that carry the flow of liquid. In a river, the

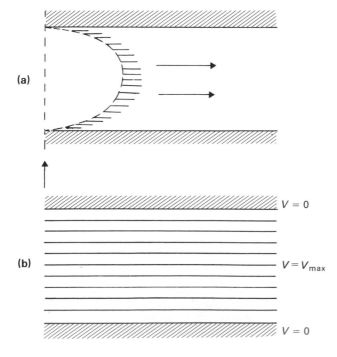

(a)

(b)

$V = 0$

$V = V_{max}$

$V = 0$

Fig. 5.5 (a) Straws dropped into a river along a line indicated by the vertical arrow map out the profile of the velocity of flow at the surface. The profile is *parabolic* in form. (b) Hypothetical streamlines of a flow at the surface of a river. The streamlines at the bank have zero velocity while the streamline in the centre has a maximal velocity.

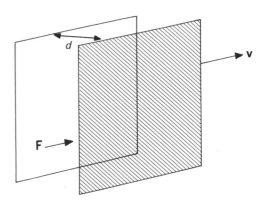

Fig. 5.6 The force **F** required to move one plate past the other in a viscous medium depends on the velocity gradient **v**/d, the area of the plates A and the viscosity of the medium.

shape of the laminae depend on the profile of the river bed, but in a cylindrical tube, the laminae would simply resemble telescope tubes. The laminae near the solid surface would scarcely move, while those at the centre would be moving with maximal velocity.

One way of determining the viscosity of a fluid is to measure the force required to maintain a certain velocity between two plates in a fluid. It is found experimentally that the force required depends on the area of the plates, the *velocity gradient* between them and of course the fluid viscosity (Fig. 5.6). The relationship is expressed in the equation

$$\mathbf{F} = \frac{A\mathbf{v}\zeta}{d} \tag{5.4}$$

where **v**/d is the velocity gradient between the two plates and ζ, the viscosity, has dimensions $m\,l^{-1}\,t^{-1}$.

Equation (5.4) applies equally well to fluid laminae and as they are considered to be infinitesimally thin, the equation takes the differential form*

$$\mathbf{F} = A\zeta\frac{d\mathbf{v}}{dx}. \tag{5.4a}$$

To find the form of the velocity profile in a system we have to integrate this equation. We shall consider the case of most use in biology, namely that of a cylindrical tube, e.g. a blood vessel.

Consider a tube of radius R (Fig. 5.7) containing a fluid flowing from left to right under the action of a pressure difference $P_1 - P_2$ ($P_1 > P_2$) across its ends a distance l apart. Fluid is carried along the tube by an infinitestinal number of laminae sliding past one another. The velocity

---

*The mathematical treatment given here is slightly more complex than that normally found in elementary texts, but I believe that it is more rigorous as it treats the case of a lamina moving in equilibrium under the action of the pressure forces rather than that of a solid cylinder of fluid (cf. Sears & Zemansky, 1964).

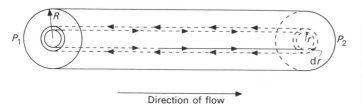

Fig. 5.7 A tube of radius R has a pressure difference $P_1 - P_2$ across its ends. The diagram shows the viscous forces on a lamina moving through the tube.

is constant within each of these laminar shells. The velocity is considered zero at the tube walls and further it is assumed to have the general profile indicated in Fig. 5.5a.

Consider further the tubular lamina of inner and outer radii r and $r + dr$ respectively. The velocity within the lamina is constant and the velocities of the fluid adjacent to the inner and outer surfaces are $\mathbf{v}_r$ and $\mathbf{v}_{r+dr}$ respectively (and $\mathbf{v}_r > \mathbf{v}_{r+dr}$). The force driving the fluid to the right along a length l of tube is the product of the pressure difference $(P_1 - P_2)$ and the cross-sectional area of the lamina ($2\pi r\, dr$). The viscous force at the outer surface of tube (F_{r+dr}) is to the left as the velocity of the lamina is greater than the velocity of the surrounding fluid. F_{r+dr} is equal to $-\zeta 2\pi(r + dr)l(d\mathbf{v}_{r+dr}/dr)$ (the negative sign occurs because $\mathbf{v}$ decreases as r increases). Similarly the frictional force ($\mathbf{F}_r$) at the inner surface is to the right and is equal to $+\zeta 2\pi rl(d\mathbf{v}_r/dr)$.

At equilibrium, the pressure driving force to the right equals the net frictional force to the left, i.e.

$$(P_1 - P_2)\, 2\pi r\, dr = -\zeta 2\pi(r + dr)\frac{l d\mathbf{v}_{r + dr}}{dr} + \zeta\, 2\pi r\frac{l d\mathbf{v}_r}{dr} \qquad (5.5)$$

but

$$\frac{d\mathbf{v}_{r + dr}}{dr} = \frac{d\mathbf{v}_r}{dr} + \frac{d^2\mathbf{v}_r}{dr^2}dr, \qquad \text{(Ferrar, 1967, chapter 7)} \qquad (5.6)$$

therefore

$$(P_1 - P_2)r\, dr = -\zeta l\frac{d\mathbf{v}_r}{dr}dr - \zeta lr\frac{d^2\mathbf{v}_r}{dr^2}dr \qquad (5.7)$$

(ignoring second order terms in dr)

$$= -\zeta l\frac{d}{dr}\left(r\frac{d\mathbf{v}}{dr}\right)dr$$

and on integration

$$\frac{(P_1 - P_2)}{2}r^2 = -\zeta lr\frac{d\mathbf{v}}{dr} + \text{constant}. \qquad (5.8)$$

Because of the general form of the profile (Fig. 5.5a) dv/dr is zero when r is zero and hence the integration constant is zero. To obtain the exact form of the velocity profile, equation (5.8) must be integrated. The limits of integration chosen are from the laminar tube to the pipe wall because we know that the velocity at the wall is zero.

$$\int_{v_r}^{0} dv = -\frac{(P_1 - P_2)}{2\zeta l} \int_{r}^{R} r\, dr \qquad (5.9)$$

and so

$$v_r = \frac{P_1 - P_2}{4\zeta l}(R^2 - r^2). \qquad (5.10)$$

This is in fact the equation of a parabola. To a biologist a picture of the velocity profile of fluid moving through a pipe is not of immediate interest, but it can be used to obtain a value for the volume flow through a vessel.

To find the volume flow across a cross-section of the tube we have to perform another integration as the velocity varies radially. We consider the volume flow across a small cross-sectional area (Fig. 5.8) a distance r from the centre where the velocity is constant and equal to v_r, i.e. across the shaded area shown in Fig. 5.8.

The volume of fluid (dQ) crossing this small area dA in time dt will be given by $v_r dA$ which is equal to $v_r 2\pi r\, dr\, dt$. If we integrate across the whole cross-sectional area, i.e. from 0 to R, we can find the rate of flow in $m^{-3}\,s^{-1}$ through the total cross-section

$$\frac{dQ}{dt} = \int_{0}^{R} v_r 2\pi r\, dr \qquad (5.11)$$

and v_r given by equation (5.10). Hence

$$\frac{dQ}{dt} = \frac{\pi(P_1 - P_2)}{2\zeta l} \int_{0}^{R} (R^2 - r^2)r\, dr \qquad (5.12)$$

$$\frac{dQ}{dt} = \frac{\pi}{8}\frac{R^4}{\zeta}\frac{P_1 - P_2}{l} \qquad (5.13)$$

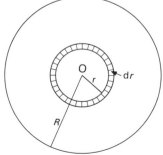

Fig. 5.8 The volume of fluid passing across the shaded area in a time dt is given by $2\pi r\, dr\, v_r\, dt$.

but

$$\frac{dQ}{dt} = A\mathbf{v} \tag{5.14}$$

where A is the cross-sectional area of the tube and $\mathbf{v}$, the mean linear velocity of flow through it. Hence

$$\mathbf{v} = \frac{R^2}{8\zeta}\frac{P_1 - P_2}{l}. \tag{5.15}$$

Equation (5.13) was first derived empirically by the physician Poiseuille who was interested in the flow of blood in the capillary bed of animals. $(P_1 - P_2)/l$ is called the pressure gradient along the tube. Although the equation predicts a linear relationship between blood flow and pressure gradient, this is not found as the blood vessels are composed of the elastic elements, elastin and collagen, which stretch under pressure, and so the vessel offers less resistance to flow than if it were rigid. Because the vessels are in a continual state of stress due to the elastic forces, there is a certain minimum pressure required to keep the tube open (Fig. 5.3).

There is also a fundamental objection to the application of the Poiseuille equation to fluids that have very large particles suspended in them. The viscosity of a fluid involves the concept of a velocity gradient measured across infinitesimally small laminae and although such a concept is accurate for a homogeneous fluid where all the particles are of molecular size, it cannot be true of blood, for example, where the laminae can take on no smaller dimensions than the thickness of the red cells. The viscosity of blood in fact depends on the concentration of the blood cells and when this is abnormally high (in polycythaemia), the viscosity of the blood can increase to as much as five times that of normal blood, imposing a similar increase in the resistance to flow of blood through the arterial system.

5.5   Reynolds number

When the velocity of a fluid flowing in a tube exceeds a certain critical value the flow is no longer laminar, but turbulent and some of the driving pressure is lost. (Fig. 5.9) Experiments have shown that a combination of four factors determine whether flow is laminar or turbulent. The combination, called Reynolds number N_R, is given by

$$N_R = \frac{\rho \mathbf{v} D}{\zeta} \tag{5.16}$$

where ρ is the density of the fluid, $\mathbf{v}$ the flow velocity, D is the diameter

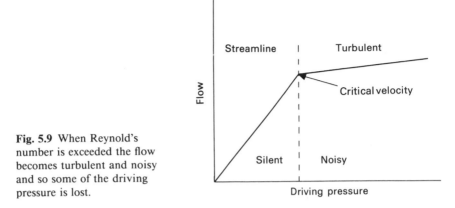

Fig. 5.9 When Reynold's number is exceeded the flow becomes turbulent and noisy and so some of the driving pressure is lost.

of the tube and ζ the viscosity of the fluid. N_R is a dimensionless quantity, and experiments have shown that when N_R is less than 2000 the flow is laminar and when N_R is greater than 3000 it is turbulent.

While the flow of blood in the aorta under normal conditions is laminar ($v = 0.3$ m s^{-1}) during heavy exercise the velocity can increase fivefold and the flow becomes turbulent (see Problem 5.5).

5.6 Methods for studying blood flow

Over the past 20 years there have been very rapid advances in the development of relatively non-invasive techniques for the study of blood flow, not only in veins and arteries, but also in particular regions of organs such as the brain and heart. The conventional, but highly accurate, invasive method that involves attaching a flow meter to a vein or artery is described in Section 9.27 (electromagnetic flow meter) and the recent, very beautiful imaging methods that require injection of radioisotopes into the body are described in Chapter 13. An additional imaging method that has the advantage that it does not require the injection of harmful isotopes is Digital Subtraction Angiography (DSA). It does, however, require the application of low doses of X-rays.

The basis for DSA is the X-ray CAT scan technique which is described in detail in Brown and Smallwood (1981). An image of the heart is first made with a digital X-ray scanner and then a contrast agent is injected into the heart via a catheter in the coronary artery. A second image is then acquired of the radio-opaque substance flowing through the heart. When the first image is substracted from the second, blood vessels containing the contrast agent can clearly be seen. Areas of blockage or constriction are highlighted (Fig. 5.10) and, in many types of DSA machines, can be imaged in three dimensions. All of the imaging techniques are in the process of development but already they

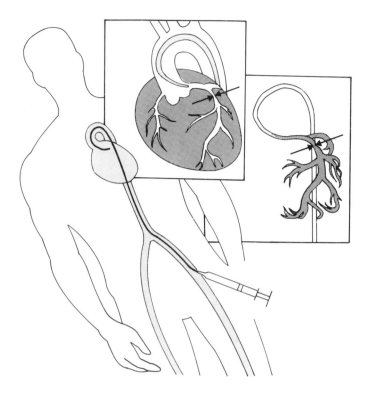

Fig. 5.10 Heart blood vessel morphology determined by Digital Subtraction Angiography (DSA). An image is first formed of the heart by digital X-ray scanning techniques (CAT scan). A contrasting agent is then injected through a catheter into the coronary arteries. A second X-ray image is made and, when the first image is subtracted by computer, only the blood vessels filled with contrast agent are left in the final computed image. Regions of vessel damage or constriction (arrowed) are clearly seen by this method (after Sochurek, 1987).

have begun to yield dynamic (i.e. rate of blood flow) as well as purely static information about the sites of particular occlusions.

One of the few techniques that does not involve the application of potentially harmful X-rays or radioisotopes, is an ultrasound, echo-location technique where Doppler analysis (Section 7.8) of the echoes from moving red cells are analysed to give information about blood flow directly. Not only is the Doppler apparatus much cheaper and safer to use, but it can give valuable information concerning the direction of blood flow as the Doppler frequency depends on the relative velocity of the source and reflector. In certain arteries, for example the suptratrochlear artery above the eyes, the direction of flow reverses when there is an occlusion of the internal carotid (Figs 5.11 and 5.12).

The normal Doppler blood flow pattern from the common carotid artery has two peaks (A and B) owing to the fact that the artery bifurcates to form the interal and external carotid arteries. The ratio of the peaks (A/B) is greater than 1 in a normal artery, but in a patient with internal carotid disease the ratio approaches 1. Similarly, the normal signal from the supratrochlear artery again shows two peaks, but in a patient with severe occlusive disease of the internal carotid the Doppler signal reveals pathological reverse flow (Fig. 5.12d).

Real-time blood flow images can be assembled using a combination of ultrasound echo-location techniques (see Section 7.10 for further

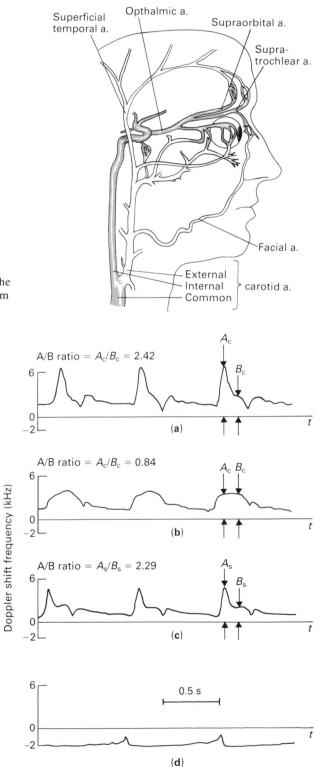

Fig. 5.11 Anatomy of the supraorbital artery (from Prichard *et al.*, 1979).

Fig. 5.12 Frequency analysed Doppler shift signals recorded from (a) the common carotid artery in a patient with normal bifurcation and extracranial vessel; (b) the common carotid artery in a patient with internal carotid disease; (c) the supratrochlear artery in a patient with normal bifurcation and extracranial vessels; (d) the supratrochlear artery in a patient with severe occlusive disease of the internal carotid showing pathological reverse flow (from Prichard *et al.*, 1979).

External and internal carotid  Common carotid

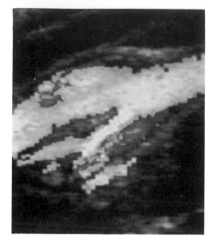 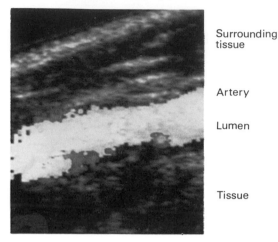

Surrounding
tissue

Artery

Lumen

Tissue

Common carotid

External carotid

Fig. 5.13 Angiodynography image of the common carotid artery and bifurcation to the external carotid. The flow pattern at peak systole in the common carotid is disturbed due to the presence of small plaques on the vessel wall.
In the bifurcated region there are large areas of turbulent flow, partly due to the normal tortuosity of the vessels involved and partly arising from plaque deposit on the vessel walls. Some considerable skill is required on the part of the operator to interpret the images obtained by this method and in modern instruments the flow velocity is colour-coded to assist interpretation (by permission Philips Medical Systems).

details). The static background details of blood vessel walls and surrounding tissues are assembled by conventional B-scan techniques while blood flow image information is computed from Doppler frequency measurements. In Fig. 5.13 the blood flow through the carotid artery is represented by shades of grey and the faster the velocity, the lighter the shade. The darker areas reveal inhomogeneities in the flow pattern and turbulent and even reversed flow can occur, especially in the external carotid.

5.7   Phloem transport

Problem 5.2

The rate of sugar transport per unit area of phloem into fruits is of the order of 0.07 mol m^{-2} s^{-1}. The average concentration of the sucrose solution moving through the phloem is 300 mM (300 mol m^{-3}).

1 Using the structural data given in Fig. 5.14, calculate the linear velocity of flow through the sieve pores and sieve lumen and hence the pressure required to drive this flow through a 1 m length of phloem, i.e. calculate the successive pressure drops due to viscosity along each section of tube lumen and sieve plate.

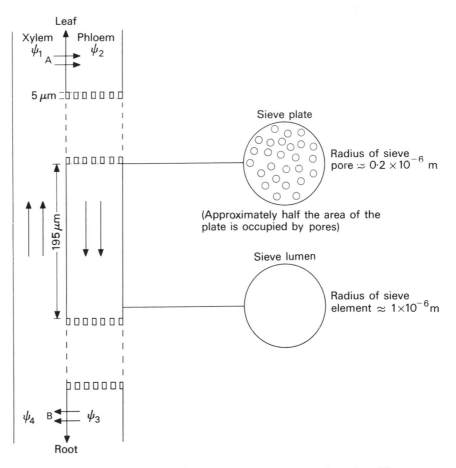

Fig. 5.14 Diagrammatic section of phloem tube and accompanying xylem. The double arrows indicate the direction of bulk flow of sucrose solution. The water potentials are denoted by ψ (Chapter 4). It has been suggested that the main driving force on the water movement is the secretion of sucrose into the upper phloem (position A) and the sucrose concentration excess is believed to be in the region of 300 mM. At A $\psi_1 > \psi_2$ because of the higher sucrose concentration in the phloem and at B $\psi_3 > \psi_4$ because of the higher hydrostatic pressure of the phloem (see also Richardson, 1969).

2 Hence show that the total osmotic driving pressure that could be developed by a difference of concentration of 300 mM sucrose across the phloem membranes, at position A, is inadequate to drive this flow. (Assume that the membranes at A are impermeable to sucrose.)

3 Can you suggest any other driving forces? A good reference to start your research into the controversial phloem transport field is Aikman & Anderson (1971) and for the physiological background see Richardson (1969). The viscosity of the sucrose solution is 2×10^{-3} N m^{-2} s.

5.8   Stokes' Law

When a sphere of radius r moves through a stationary fluid and if the motion is non-turbulent, the viscous drag F on the sphere is given by Stokes' Law

$$F = 6\pi \zeta r \mathbf{v}. \tag{5.17}$$

When such a sphere falls in a viscous medium it reaches a terminal velocity when the retarding forces, viscosity and buoyancy, equal the weight of the sphere. The weight of the sphere is $\frac{4}{3}\pi r^3 \rho_s \mathbf{g}$ and the buoyancy force equal to the weight of fluid displaced, is $\frac{4}{3}\pi r^3 \rho_f \mathbf{g}$ where ρ_s and ρ_f are the densities of the sphere and fluid respectively. When the terminal velocity $\mathbf{v}_T$ has been reached

$$\frac{4}{3}\pi r^3 \rho_s \mathbf{g} = \frac{4}{3}\pi r^3 \rho_f \mathbf{g} + 6\pi \zeta r \mathbf{v}_T$$

hence

$$\mathbf{v}_T = 2r^2(\rho_s - \rho_f)\mathbf{g}/9\zeta. \tag{5.18}$$

The terminal velocity is also called the *sedimentation velocity* when applied to the centrifugation of macromolecules (Fig. 5.15)

$$\mathbf{v}_T = \omega^2 R \, 2 \, r^2 \, (\rho_s - \rho_f)/9\zeta \tag{5.19}$$

where ω is the angular velocity of the centrifuge tube and R is the distance of the molecule from the axis of rotation. The fact that R appears in the equation shows that it is meaningless to quote rpm alone when quoting centrifuge data; in fact, the number of **g**s in $\omega^2 R$ is the

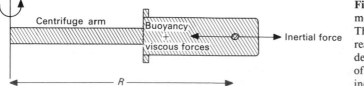

Fig. 5.15 Movement of a molecule in a centrifuge tube. The molecule eventually reaches a terminal velocity, $\mathbf{v}_T$, determined by the resultant of the buoyancy, viscous, and inertial forces.

usual means of indicating the conditions under which macromolecules are sedimented.

The sedimentation velocity of human erythrocytes increases greatly in certain medical conditions (e.g. pregnancy) and this is due to the aggregation of the corpuscles into larger sedimentary masses.

Problem 5.3

(a) The mean volume flow of blood in the arterial system is 8.5×10^{-5} $m^3\,s^{-1}$. If the area of the aorta is $3 \times 10^{-4}\,m^2$, find the linear velocity of flow and show that the average kinetic energy per unit volume is approximately $45\,J\,m^{-3}$. Show that the pressure corresponding to this kinetic energy per unit volume is only a small fraction of the overall arterial pressure.

(b) At the beginning of the heart's ejection period, the linear velocity of the blood may be three times the average velocity. Show that this explains the experimentally measured pressure difference of some 200 $N\,m^{-2}$ between the *aorta*, into which the blood is rushing, and the *left ventricle*, where the velocity is near zero, during this short period.

(c) Show that during heavy exercise, when the cardiac output is five times normal, the kinetic energy created becomes a significant fraction of the total work done by the heart.

Problem 5.4

In diseased arteries, an *atherosclerotic plaque* can narrow the lumen to under one-fifth of the original area (Fig. 5.3). By how much does this reduce the arterial pressure within the constricted area?

Problem 5.5

(a) The resting velocity of flow through the aorta of diameter $2 \times 10^{-2}\,m$ is $0.3\,m\,s^{-1}$. If the density and viscosity of blood are $10^3\,kg\,m^{-3}$ and $4 \times 10^{-3}\,kg\,m^{-1}\,s^{-1}$ respectively, show that the flow is laminar.

(b) Show that during heavy exercise this need not necessarily be so (see Problem 5.2).

References

Aikman D.P. & Anderson W.P. (1971) A quantitative investigation of a peristaltic model of phloem translocation. *Annals of Botany* **35**, 761–772.

Brown, B.H. & Smallwood, R.H. (1981) *Medical Physics and Physiological Measurement*. Blackwell Scientific Publications, Oxford.

Ferrar W.L. (1967) *Calculus for Beginners*. Clarendon Press, Oxford.

Prichard, D.R., Martin, T.R.P. & Sherriff, S.B. (1979) *Journal of Neurology, Neurosurgery and Psychiatry*, **42**, 563–8.
Richardson, M. (1969) *Translocation in Plants*. Arnold, London.
Sears, F.W. & Zemansky, M.W. (1964) *Unversity Physics*. Addison–Wesley, Reading, Mass.
Sochurek, H. (1987) Medicine's new vision. *National Geographic*, **171**, 2–41.

Further reading

Alexander R.M. (1968) *Animal Mechanics*. Sidgwick & Jackson, London.
Alexander R.M. (1971) *Size and Shape*. Arnold, London.
Barton, A.C. (1966). *Physiology and Biophysics of the Circulation*. Year Book Medical Publishers, Chicago.
Dendy, P.P. & Heaton, B. (1987) *Physics for Radiologists*. Blackwell Scientific Publications, Oxford.
McDonald, D.A. (1974) *Blood Flow in Arteries*. Edward Arnold, London.
Merritt, C.R.B. (1986) Doppler blood flow imaging: integrating flow with tissue data. *Diagnostic Imaging* **8**, 146–55.
Randall, J. E. (1959) *Elements of Biophysics*. Year Book Medical Publishers, Chicago.
Ruch, T.C. & Patton, H.D. (1965) *Physiology and Biophysics*. Saunders, Philadelphia.
Setlow, R.B. & Pollard, E.C. (1962) *Molecular Biophysics*. Pergamon, London.

6 Surface Tension

6.1 Introduction and definitions

In a homogeneous fluid, the molecules in the body of the fluid have on average no net forces acting on them as their nearest neighbours pull equally in all directions. At the surface of the fluid, however, there is a net force on a molecule which acts towards the bulk phase (Fig. 6.1). The net effect of these inward pulling forces is called the *surface tension*.

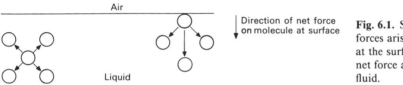

Air

Liquid

Direction of net force on molecule at surface

Fig. 6.1. Surface tension forces arise because a molecule at the surface experiences a net force acting towards the bulk fluid.

When a wire ring carrying a loop of thread is dipped in a soap solution and withdrawn, the soap forms a film on the ring and the thread lies slack in the film (Fig. 6.2a). If, however, the film within the thread is punctured the loop is pulled taut by the action of the soap molecules in the bulk phase (Fig. 6.2b).

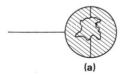

(a)

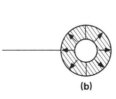

(b)

Fig. 6.2. (a) When the soap film completely covers the loop the thread lies slack in the film. (b) When the film in the centre is broken, however, the thread is pulled taut by surface tension forces.

The forces involved in surface phenomena at a soap–air interface can be measured by means of a slider of mass m_1 and length l which can move easily up and down the arms of a wire bent in the shape of a U (Fig. 6.3). In this device there are two interfaces along which surface tension forces act on the slider, i.e. on the front and back surfaces (Fig. 6.3b). When the wire is dipped in solution with the slider at the bottom, it is rapidly pulled to the top under the action of the surface tension forces.

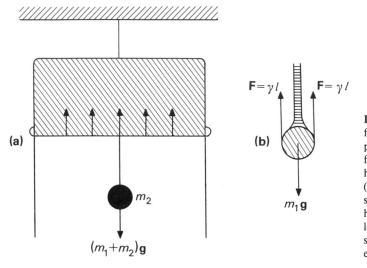

(a)

(b)

$F = \gamma l$ $F = \gamma l$

m_2

$(m_1 + m_2)\mathbf{g}$

$m_1\mathbf{g}$

Fig. 6.3. (a) Surface tension forces at the soap–air interface pull the slider up and the forces can be measured by hanging weights on the slider. (b) Cross-section through wire slider and soap film. The film has two interfaces, each of length l and hence the total surface tension force upwards equals $2\gamma l$.

The slider can be held in equilibrium by attaching a small weight m_2, and it is found that this weight will hold the slider at any position on the wire. This means that the surface forces, which are equal to $(m_1 + m_2)\,\mathbf{g}$ are independent of the total area of the film. Hence surface phenomena differ from elastic ones as, in the latter case, stretching a body requires greater forces to maintain equilibrium (Chapter 2). In surface phenomena the molecules are not stretched, but the surface in fact expands at the expense of molecules from the bulk phase.

The surface tension forces depend only on the molecular species on either side of the interface, and on the length of the interfaces. The total length of the soap–air interface is $2l$ (Fig. 6.2b) and the surface tension γ is defined as the force acting perpendicular to the interface, divided by the total length of the interface, i.e.

$$\gamma = \frac{(m_1 + m_2)\,\mathbf{g}}{2l} = \frac{\mathbf{F}}{2l} \tag{6.1}$$

and surface tension has the units $N\ m^{-1}$. Table 6.1 gives the surface tension of some common liquids when they are in contact with air.

Table 6.1. (*See text*)

Liquid in contact with air	$t°C$	Surface tension ($\times\ 10^3$) ($N\ m^{-1}$)
Benzene	20	29
Soap solution	20	about 25
Bile-salt solution	20	about 20
Water	20	72

6.2 Surface energy

Although no extra force is required to expand the area of the soap film, work is done when the area is increased. If the slider is moved a distance y, the work done is $\mathbf{F}y$ and the total increase in surface area is $2ly$. The work done per unit area in expanding the film is therefore

$$\mathbf{F}y/2ly = \mathbf{F}/2l = \gamma. \tag{6.2}$$

Hence the surface tension is also the *potential energy stored in a unit area of a surface.*

In the same way as mechanical systems tend to move to a position of minimum potential energy and chemical systems move to a position of minimum chemical energy, surfaces seek an arrangement of minimal surface energy. In a solution this can be attained by the formation of a mass of minimum surface area, such as a sphere. Soaps, lipids, and bile salts, for example, tend to lower the surface tension at an air–water interface and so they will tend to accumulate at the interface because their presence lowers the surface energy.

Bile salts aid the digestion of fats by lowering the surface tension at the fat–water interface. This permits an increase in the total area of interface and so one large fat globule can break up into several small ones allowing for a greater area of attack by pancreatic lipases in the small intestine.

6.3 Contact angle

In many situations there are three phases where tension forces interact, for example in the rise of a water meniscus in contact with glass (Fig. 6.4). The liquid–air interface lies at an angle θ to the glass wall, and θ is called the *contact angle*. When θ is less than 90°, then the liquid is said to *wet* the solid and this is the case illustrated. The contact angle between mercury and glass is greater than 90° and so when a capillary tube is pushed into mercury the surface within the tube is lower than that outside. The contact angle between pure water and paraffin wax is 110°, so water does not wet paraffin wax. If detergents are added to the water,

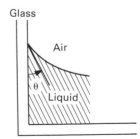

Fig. 6.4. At a water–air interface in a glass vessel the contact angle is less than 90° and so the water is said to wet the glass.

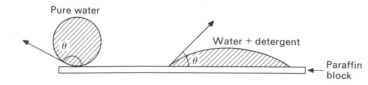

Fig. 6.5. If a detergent is added to the water, it will wet the paraffin.

the contact angle is reduced and the water will then wet the paraffin (Fig. 6.5).

6.4   The Laplace equation

Very often we have to calculate the forces acting on spheres or hemispheres in contact with air. Consider the surface tension forces on a bubble of air in a liquid. The right half of the sphere is shown in the diagram (Fig. 6.6) and it is attracted to the left by surface tension forces

$$\mathbf{F} = \gamma l = \gamma 2\pi r$$

where r is the radius of the sphere. The force acting in the opposite direction is derived from the pressure difference across the bubble, and the component acting anti-parallel to the surface tension forces is the pressure difference acting on the projected area πr^2 (Fig. 6.6).

The total force to the right is $(P_1 - P_2)\,\pi r^2$. Equating the two forces

$$(P_1 - P_2)\,\pi r^2 = 2\pi r\gamma \tag{6.3}$$

$$P_1 - P_2 = \frac{2}{r}\gamma \tag{6.4}$$

or $$\Delta P = \frac{2}{r}\gamma.$$

The collapse pressure on the bubble, i.e. pressure ΔP required to keep the bubble in equilibrium, increases as the radius decreases.

Laplace's equation can be used to calculate the rise of liquids in narrow tubes (Fig. 6.7). At a glass–air interface the contact angle is near zero and so the meniscus can be considered as tangential at the glass and

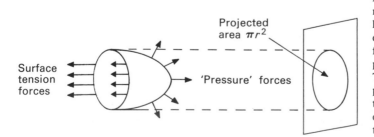

Projected area πr^2

Surface tension forces

'Pressure' forces

Fig. 6.6. The equilibrium radius of a bubble of air in a liquid can be computed by equating the surface tension forces (to the left) and the pressure forces (to the right). The plane on which the projected area falls is parallel to the 'cut' surface used for the computation of the surface tension forces.

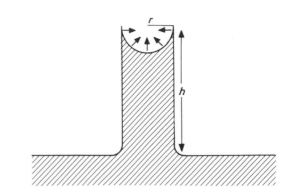

Fig. 6.7. The fluid in a narrow tube is pulled up by collapse pressure on the hemisphere at the liquid–air interface.

hemispherical in the tube lumen. The collapse pressure serves to drag a column of water upwards as it is not balanced by interactions with an opposing hemisphere.

The upward force will be the collapse pressure × the projected area and this will be balanced by the downward force due to the action of gravity on the fluid mass, i.e.

$$\frac{2\gamma\pi r^2}{r} = m\mathbf{g}. \tag{6.5}$$

Now $m = \rho V = \rho\pi r^2 h$; where V is the volume of the fluid, ρ the density; and h the height to which it rises. Hence

$$h = \frac{2\gamma}{\rho r \mathbf{g}}. \tag{6.6a}$$

Equation (6.6), as derived, is only strictly true in the case where the contact angle (Fig. 6.4) at the liquid–air interface is zero. In the general case when the contact angle is θ, then equation (6.6a) becomes

$$\rho r \mathbf{g} h = 2\gamma \cos\theta. \tag{6.6b}$$

Problem 6.1

Sap flows upwards through the xylem in trees. Diameters of tracheids and vessels range from 20 to 400 μm. Can capillarity alone account for the sap rising to a height of 100 m in some trees, e.g. *Sequoia*? If not, suggest other mechanisms by which this height might be reached (see Richardson, 1969). Take density and surface tension of sap as 10^3 kg m^{-3} and 70×10^{-3} N m^{-1} respectively.

6.5   Surface balance

The instrument used in biophysics laboratories to measure surface tension quantitatively is the surface balance (Fig. 6.8). It consists of a

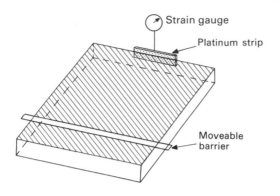

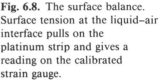

Fig. 6.8. The surface balance. Surface tension at the liquid–air interface pulls on the platinum strip and gives a reading on the calibrated strain gauge.

water-filled Teflon® trough with a movable Teflon® barrier. At one end a platinum strip dips in the water and it is attached to a calibrated strain gauge which gives a direct measure of the surface tension force pulling down on the strip. Water alone produces a pull of $70 \times 10^{-3} \, \text{N m}^{-1}$. When detergent is added the surface tension is reduced to about $30 \times 10^{-3} \, \text{N m}^{-1}$ and the tension does not change as the barrier moves back and forth. When phospholipids, e.g. phosphatidyl choline, are added to pure water the surface tension is reduced but in this case, as the barrier is moved towards the platinum strip, the surface tension is reduced still further. This is because phospholipids are relatively water insoluble and are concentrated at the available surface. Their charged groups are water soluble and so remain in contact with the water, but their fatty acid chains are hydrophobic entities and in fact wave about in the air above the surface. As the barrier sweeps towards the strip the phospholipid molecules are forced closer and closer together and the surface tension is progressively lowered. In fact if a known amount of phospholipid is added, the surface balance will give a measure of the area per molecule when the barrier has been moved up to a position of minimum area (M in Fig. 6.10). As the energy to bring the molecules together to a certain configuration is not the same as that required to separate them, the surface tension–area curve for inward and outward

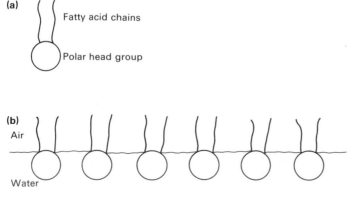

Fig. 6.9. (a) Phospholipids are surface active species; they consist essentially of a polar group and one or more fatty acid chains. (b) Because of the hydrophobic nature of the fatty acid chains, phospholipids gather at the water–air interface, and are oriented so that the polar groups make contact with the water and the fatty acid chains stick up into the air. (See also Chapter 9).

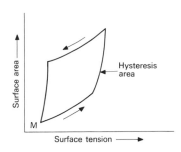

Fig. 6.10. The surface tension encloses a hysteresis curve.

sweeps of the barrier encloses a hysteresis curve (Fig. 6.10), i.e. net work has to be done on the system during each cycle.

6.6   Surface tension and the lung

One of the most important interfaces to terrestrial animals is the blood–epithelium–air interface in the lungs. In order to promote gaseous interchange, the area of this interface is very large and in man it is approximately equal to that of a tennis court. In order to fit the vast lung area into the chest, the lung itself is highly invaginated (Fig. 6.11), and the tiny air spaces which are almost spherical in shape are called *alveoli*. The normal functioning of the lung demands a high concentration of certain surface active species in the walls of the alveoli.

Karl von Neergaard, a pioneer investigator of lung mechanics, first demonstrated the important role of surface tension by some ingenious and simple experiments (Fig. 6.12). He distended lungs first with air and then with saline and found that more pressure was required to expand in air than in saline (Fig. 6.13). In the first case both surface tension and elastic forces have to be overcome whereas in the latter there

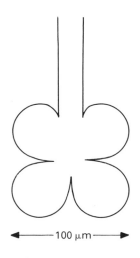

Fig. 6.11. In order to fit the vast surface area of the lung into the chest it is highly invaginated, and the figure shows a diagrammatic cross-section of a group of alveoli.

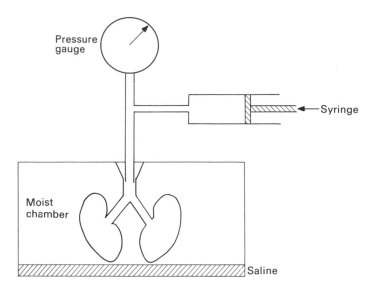

Fig. 6.12. Von Neergaard's apparatus to measure surface tension contribution to lung mechanics. The excised lungs were cannulated and connected to a syringe and pressure gauge. In this way the lung could be filled with either air or saline (after Clements, 1962). Copyright (1962) by Scientific American Inc. All rights reserved.

are only elastic forces. Von Neergaard obtained the surface tension contribution from the difference between the two pressure–volume curves.

Von Neergaard's technique has been used to demonstrate that the surface tension contribution is much higher in a fatal respiratory disease in new born babies, called the hyaline membrane disease, than it is in the normal lung. This finding led to a search for a surface active species that was reducing surface tension at the blood–air interface in the normal lung but was absent in diseased lungs. It is only relatively recently, however, with the introduction of sophisticated biochemical techniques that much progress has been made. The introduction of the surface balance (Fig. 6.8) into lung research was also required before the reason for the importance of surface tension was properly understood.

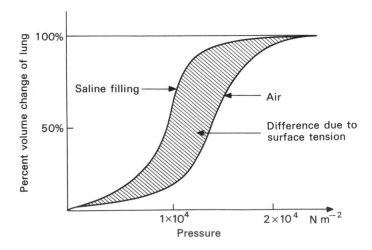

Fig. 6.13. Pressure–volume relationships when lung is filled with air or saline (after Clements, 1962).

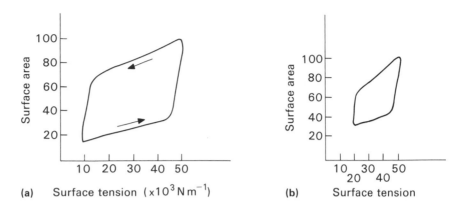

Fig. 6.14. (a) Data from lung extract in surface balance apparatus show a hysteresis curve typical of phospholipids. At the minimum area, the surface tension is only about 8×10^{-3} N m^{-1}. (b) Surface tension data from lung extract of a baby that had died from hyaloid membrane disease. At the minimum surface area the surface tension is still over 20×10^{-3} N m^{-1} (after Clements, 1962).

When an extract of normal lung is added to the water in the balance there is immediately a small but significant reduction in the surface tension (Fig. 6.14) from 70 to 40×10^{-3} N m^{-1}. However there is a spectacular decrease when the barrier is moved towards the platinum strip and when a position of minimum area is reached the surface tension has been reduced to below 10×10^{-3} N m^{-1}. The shape of the area versus surface tension curve is typical of that for phospholipids (Fig. 6.10) and a relatively high concentration of phosphatidyl choline (Fig. 6.15) can be extracted from the normal lung where it seems to be anchored to the interfacial membranes by a low molecular weight protein. The surface tension from the diseased lung on the other hand does not fall below 20×10^{-3} N m^{-1} (Fig. 6.14b).

An estimation of the total possible collapse pressure on an alveolus, (of average radius 5×10^{-5} m), can be found by applying the Laplace equation (6.4)

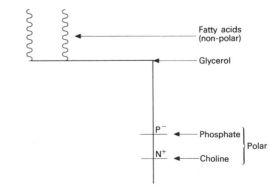

Fig. 6.15. Diagram of lecithin (phosphatidyl choline). (See also Chapter 9 for the arrangement of phospholids in membranes).

$$\Delta P = \frac{2\gamma}{r}$$

and as $\gamma = 50 \times 10^{-3}\,\text{N m}^{-1}$ for plasma, a value of $2 \times 10^3\,\text{N m}^{-2}$ is obtained for ΔP. This is in fact a considerable pressure acting to collapse the lungs and explains why the absence of a surface active agent leads to respiratory failure.

Laplace's equation also demonstrates the importance to the lung of the area dependence of surface tension. A decrease in r might be expected to lead to an increase in the collapse pressure. However, this increase is offset by a decrease in the surface tension following a compression of the surface area. Hence a relatively homogeneous distribution of pressure is brought about by the surface active species and the performance of the many alveoli, of random size, is smoothed and coordinated.

6.7 Importance of surface tension kinetics in respiratory distress

Since the importance of lung surfactants in respiratory distress syndrome in premature babies has been largely understood for over 20 years, the lack of success in using surfactant sprays, nebulized in water, to treat the disease has been very surprising. However, recent experiments have shown that other forces must play a role in preventing alveolar collapse during exhalation. A more complete understanding of the forces involved has led to exciting new developments in the production of artificial surfactants to overcome hyaline membrane disease. The first clues to the existence of additional mechanisms came from the fact that surfactant sprays of phospholipids nebulized in water were relatively ineffective in the treatment of the disease. 'Dry' sprays

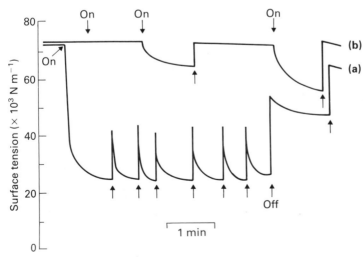

Fig. 6.16. The surface tension lowering properties of (a) a dry surfactant and (b) a wet surfactant, placed on saline. In both instances the surfactant was placed on the surface at 'On'. The dry particle was removed from the surface at 'Off'. Upwards-pointing arrows indicate when the surface monolayer was partly removed by aspiration. (From Morley *et al.*, 1978).

of mixtures of dipalmitoyl phosphatidylcholine (DPPC) and phosphatidyl glycerol (PG), however, are extremely effective (Bangham, 1987). Some recent, simple but extremely elegant experiments demonstrate some of the reasons for their success.

Surfactant was first extracted from the lungs of premature lambs and, after purification, was evaporated to dryness. In one set of experiments the surfactant was used in the dry state and, in the other, it was thoroughly mixed with saline solution (of the same NaCl content as blood). When a small particle of dry surfactant was placed on a clean interface of a surface balance apparatus, the surface tension fell rapidly to the equilibrium value of $24 \times 10^{-3} \, \text{N m}^{-1}$ (Fig. 6.16). Even when the surface film was removed on aspirating with a clean Pasteur pipette, the surface tension returned to the equilibrium value after an initial rise. It was only when the particle itself was removed that the surface tension rose irreversibly on aspiration.

Interestingly, the surfactant was quite ineffective when applied to the surface pre-mixed in saline. It appears that the molecular configurations of the dry and wet surfactants are quite different and it is only in the dry, lamellar phase, that surfactant molecules are free to leave the particle to form a monolayer. The closed form of the wet 'smectic mesophase' (liposome) precludes the ready exit of individual molecules (Fig. 6.17). In this phase the concentration of free phospholipids is given by the critical micellar concentration (CMC) and is less than 10^{-10} M.

The composition of the artificial surfactant spray is extremely important and DPPC and PG are vital components. DPPC is an essential component because a surface monolayer of DDPC is in-

Fig. 6.17. The molecular configuration of dry and wet surfactant. (a) In water the surfactant aggregates as a smectic mesophase and few molecules are free to reach the interface. (b) The dry surfactant has open-ended layers from which molecules can spread freely to an air–water interface (from Morley *et al.*, 1978).

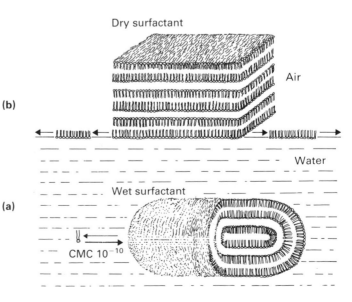

compressible at 37°C and it is now believed that this incompressibility prevents alveolar collapse during exhalation. DPPC on its own, however, does not spread easily, so during the exhalation cycle other surface active species such as PG, with good spreading characteristics, have to be added.

It is therefore necessary, to understand that alveolar contraction and expansion represents a kinetic system, with phospholipids entering and leaving the surface during the cycle. They are only able to enter from the 'dry' configuration and merely represent an inactive reservoir when encapsulated as liposomes. During that first, most important breath of all, 20 m^2 of lung–air interface has to be formed and hence rapid recruitment of the surface active molecules from an accessible store is vital.

Problem 6.2 (after Jarman, 1970)

An insect called the pond skater can walk on water. It has six feet and the total length of the air–water interface at each foot has been estimated to be 1 mm.

1 Assuming that the contact angle between the foot and water is sufficiently large so that the surface tension ($\gamma = 70 \times 10^{-3}\,\text{N m}^{-1}$) acts vertically, show that this force alone can support the pond skater of mass 25×10^{-6} kg.

2 Show that the skater will sink when the surface tension is lowered to $40 \times 10^{-3}\,\text{N m}^{-1}$. (This can be rather unkindly demonstrated using detergents.)

References

Bangham, A.D. (1987) Lung surfactant: how it does and does not work. *Lung*, **165**, 17–25.
Clements, J.A. (1962) Surface tension in the lungs. *Scientific American,* **207**, 6, 121–30.
Jarman, M. (1970) *Examples in Quantitative Zoology*. Arnold, London.
Morley, C.J., Bangham, A.D., Johnson, P., Thorburn, G.D. & Jenkin, G. (1978) Physical and physiological properties of dry lung surfactant. *Nature*, **271**, 162–3.
Richardson, M. (1969) *Translocation in Plants*. Arnold, London.

Further reading

Alexander, R.M. (1968) *Animal Mechanics*. Sidgwick & Jackson, London.
Alexander, R.M. (1971) *Size and Shape*. Arnold, London.
Clements, J.A., Brown, E.S. & Johnson, R.P. (1958) Pulmonary surface tension and the mucous lining of the lungs: some theoretical considerations. *Journal of Applied Physiology*, **12**, 262–6.
Sears, F.W. & Zemansky, M.M. (1964) *University Physics*. Addison-Wesley, Reading, Mass.
Zimmermann, M.H. (1963) How sap moves in trees. In *From Cell to Organism*. Freeman, San Francisco.

7 Sound and Ultrasonics*

7.1 Introduction

The sense of hearing plays a large part in our lives; the spoken word enables us to communicate and through listening to music we both relax and derive intellectual satisfaction. Sounds also have great significance for other animals: bats use high frequency sounds for navigation and for the catching of food; insects have very loud courtship songs; dolphins are well-known for their ability to communicate with their fellows by sound alone; and whales can in fact communicate over hundreds of miles by emitting coded, low frequency sounds.

Sound energy travels from source to receiver in the form of waves and there are certain similarities between sound and light waves. Both can be reflected or refracted at the interface between media of different compositions and both can produce *interference* and *diffraction patterns* (see Chapter 8 for an explanation of these terms). However, there are important differences.

7.2 Differences between sound and light waves

1 Sound energy is transmitted from one point to the next by vibrations of the molecules comprising the medium between the points. This implies that sound waves cannot travel *in vacuo*, an experimental fact, whereas light waves can.

2 Sound is a transfer of mechanical energy, whereas light is a transfer of electromagnetic energy.

3 Light waves are *transverse* vibrations whereas sound waves, in a gas, are *longitudinal* vibrations. In longitudinal wave motion the vibrations of the molecules in the medium are along the direction of travel of the wave. One analogy for a longitudinal wave is an arrangement of heavy balls connected by springs (Fig. 7.1a). If the first ball is set in motion by means of a horizontally oscillating force **F** (Fig. 7.1b) then a longitudinal wave will be sent out towards the right as the balls will alternatively bunch together and spread out giving regions of compression C and rarefaction R.

---

*Readers who have not met the concept of wave motion before should postpone reading this chapter until after Chapter 8.

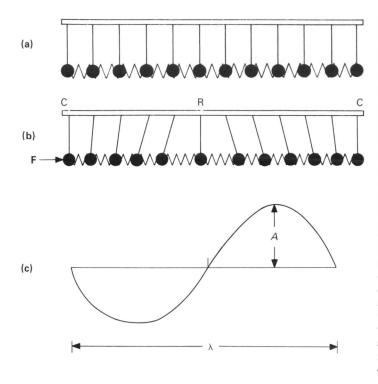

Fig. 7.1. (a) An analogy for a longitudinal wave travelling in an elastic medium. The balls represent the molecules comprising the medium and the links between them represent the compressibility. (b) An oscillating force **F** acting to the left will result in a longitudinal force being sent out towards the right. This consists of alternating regions of compression C and rarefaction R in the medium. (c) *A* is the amplitude of the disturbance. The wavelength of the disturbance is λ and this is the distance between the successive regions of compression (or rarefaction). If c is the velocity of a wave, and f is the frequency, i.e. number of cycles per second, then λ, the distance travelled by a region of compression (or rarefaction) during one cycle, equals c/f.

If the displacements of each of the balls from the horizontal are plotted, left displacements downwards, against the distance of the ball from the origin, the wave-like nature of the motion is apparent (Fig. 7.1c). Longitudinal waves can therefore be transmitted through any medium that offers elastic resistance, but the present treatment will be restricted to fluids.

4   The velocity of sound waves *increase* on travelling from air to water, whereas the reverse is true for light waves. The velocity of sound in air is about 330 m s^{-1} whereas in water the velocity is about 1500 m s^{-1}. From the relationship

$$\mathbf{c} = f\lambda \tag{7.1}$$

where **c** is the velocity, f the frequency and λ the wavelength of the sound wave, which holds for longitudinal as well as transverse waves (Fig. 7.1c), it follows that the wavelength of a sound wave of a certain given frequency increases more than fourfold when the wave travels from air to water.

7.3   Physical characteristics of sound waves

The sounds that we hear have three characteristics: (i) *pitch*, (ii) *loudness*, and (iii) *tone quality* corresponding to the three physical quantities of *frequency, intensity,* and *waveform.*

1 The *pitch* of a note is determined by the frequency — the more vibrations per second (Hz) of the sound source, the higher will be the pitch of the note and, in the musical scale, doubling the frequency of a note raises its pitch one *octave*.

The human ear is able to hear sounds in range 20 to 20 000 Hz and frequencies above this are said to be ultrasonic.

Any body capable of oscillating can emit sound waves and the frequency spectrum of most bodies is complex as a range of frequencies can be emitted. However, if such a body is placed within range of a tuning fork emitting a note of frequency f, and if this is one of the natural frequencies of the body, the body too will begin to vibrate and emit *sound* of the same frequency. This phenomenon is termed *resonance*. (See also Section 2.20)

2 The *intensity* or *loudness* of a sound is a measure of the energy impinging on unit area of receiver surface in unit time. The units of intensity are therefore W m^{-2}.

Consider the sound waves set up by a piston vibrating in an open-ended tube (Fig. 7.2). The sound wave is set up initially by the piston moving towards the right with velocity **v**. The piston moves through a distance **v**t in a time t and this sets in motion a column of air of length **c**t where **c** is the velocity of sound in air. P is the excess pressure, i.e. the pressure over and above the atmospheric pressure, which has to be applied to the piston to accelerate the mass of air originally contained in the column **c**t. The individual gas molecules in the column in fact are displaced with a velocity **v**, whereas the *wavefront* of the disturbance moves with a velocity **c**. As

$$\mathbf{F} = m\mathbf{a} \tag{7.2a}$$

then

$$PA = \rho\,\mathbf{c}t\,A\mathbf{a} \tag{7.2b}$$

where ρ is the density of air, and A is the cross-sectional area of the pipe. As the individual gas molecules are accelerated from rest to a velocity **v** in time t,

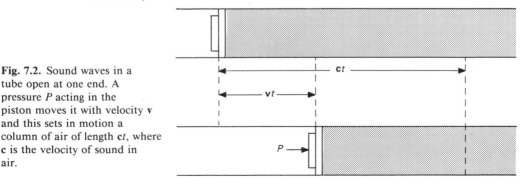

Fig. 7.2. Sound waves in a tube open at one end. A pressure P acting in the piston moves it with velocity **v** and this sets in motion a column of air of length **c**t, where **c** is the velocity of sound in air.

$$PA = \rho\, \mathbf{c}t\, A\, \mathbf{v}/t \qquad\qquad (7.3)$$

$$\text{or}\quad P = \rho\, \mathbf{cv}, \qquad\qquad (7.4)$$

$$\text{i.e.}\ P = \text{constant} \times \mathbf{v}. \qquad\qquad (7.4a)$$

The constant in equation (7.4) (ρc) defines the *acoustic impedance* of the medium. Note that it gives the relationship between pressure (a scalar quantity) and particle velocity (a vector).

The acoustic impedance of a *system* on the other hand, for example the auricle of the ear with an opening of cross-sectional area A, is defined as the relationship

$$\text{impedance} = \frac{P}{A\mathbf{v}} \quad \text{(von Békésy, 1960).} \qquad (7.4b)$$

The acoustic impedance of air is 430 kg m^{-2} s^{-1} and it can be shown that if there is not to be a serious energy reflection when sound travels from one medium to another, e.g. from the air to tissues of the ear, the acoustic impedances of the two must be matched (see Problem 7.1).

The power carried per unit area of a column (I) is given by the force per unit area multiplied by the velocity, i.e.

$$I = \frac{\mathbf{F}\mathbf{v}}{A} \qquad\qquad (7.5)$$

$$= \rho\, \mathbf{c}\, \mathbf{v}^2. \qquad\qquad (7.6)$$

This is the instantaneous power or intensity and it has units W m^{-2}. Normally we are concerned with the average intensity that is carried over one cycle of a sinusoidally varying sound wave and in this case it can be shown that

$$\text{average intensity } I_{av} = \tfrac{1}{2}\rho\mathbf{c}\,\mathbf{v}^2 \qquad (7.7)$$

$$= \tfrac{1}{2}\frac{P_0{}^2}{\rho\mathbf{c}} \qquad\qquad (7.8)$$

when the excess pressure on the piston at any time t is given by

$$P = P_0 \sin 2\pi\ ft. \qquad\qquad (7.9)$$

The intensity of the faintest sound which can just be heard is about 10^{-12} W m^{-2} which corresponds to a pressure amplitude P_0 of about 3×10^{-5} N m^{-2}. The loudest tolerable sound has an intensity of approximately 1 W m^{-2} and a pressure amplitude of 30 N m^{-2}.

Because of this wide range in intensities over which the ear operates, and because the ear can just discriminate between sounds of a certain intensity ratio whether they are loud or soft, a logarithmic rather than a linear intensity scale is used. The *intensity level B* of a sound wave is defined by the equation

$$B = \log_{10} \frac{I}{I_0}, \tag{7.10a}$$

where I_0 is an arbitrary reference intensity and is conventionally taken as the threshold of hearing, 10^{-12} W m^{-2}.

The intensity level is a dimensionless quantity and the unit is the *bel* (B) in honour of A. G. Bell. In practise a unit of 0.1 B or *decibel* (dB) is more frequently used. This unit may also be used for describing the attenuation or amplification of sound or even electrical signals. Thus an amplifier with a gain of 1000 (Chapter 11) can also be described as having a gain of 30 dB. The intensity level of the threshold of hearing is taken as 0 dB and the loudest tolerable level is then 120 dB.

From equation (7.8) the decibel can also be described as

$$dB = 20 \log_{10} \frac{pressure}{threshold\ pressure}. \tag{7.10b}$$

When the sound consists of a mixture of frequencies, as sounds generally do, then the weighting given to each frequency must be stated. There are internationally agreed weighting schemes, for instance the A weighting which corresponds to the sensitivity of the ear and the D weighting used in aircraft noise measurements. The weighting used is put in brackets after dB.

3 The *tone quality* corresponds to the complexity of the waveform produced by the source. A well-made tuning fork will vibrate sinusoidally with only one frequency and so produced a pure tone. However, experience shows that most vibrating bodies, in addition to the *fundamental* or lowest frequency, have *harmonics* which are frequencies that are integral multiples of the fundamental. If f is the fundamental frequency, then $2f$ is called the second harmonic. The number and relative intensity of the harmonics determine the tone quality of the note. The oscillogram or graph of the waveform from the note of a violin played on the open G string is shown in Fig. 7.3 (a). The relative amplitudes of the various harmonics that make up the waveform are given as the *harmonic analysis* (Fig. 7.3b) and *frequency spectrum*. Figure 7.3 (c) is in fact the result of carrying out a *Fourier analysis* of the waveform and is a representation in the frequency domain of an event that occurs in the time domain. A surprising observation is that the fundamental frequency (196 Hz) is missing and this is because the body of the violin does not resonate at such a low frequency. However, even more surprising is the fact that the quality of the note is unchanged even

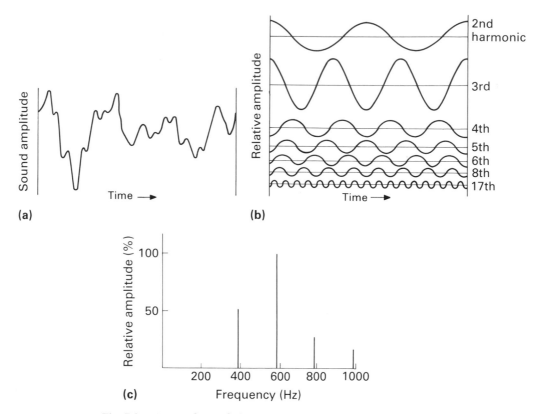

Fig. 7.3. (a) Waveform of violin tone G = 196 Hz. (b) Waveform analysed into its harmonics. (c) The frequency spectrum of a violin G string. Note that the fundamental (196 Hz) is missing.

when the fundamental is missing. It is said that the ear supplies the fundamental as this is the common *difference* tone of the upper harmonics. This fact is used to good advantage in cheap loudspeakers which give a reasonable reproduction to the untrained ear even if they are incapable of producing low frequencies.

7.4 Beats

When two tuning forks emitting different frequencies are sounded together, a periodic rise and fall in the intensity of the sound can be perceived arising from the interference of the sound waves from the two sources. The phenomenon is called *beats* and the number of beats per second is equal to the difference in frequencies of the two forks.

Consider the rotating vector representation of the waves (Section 2.10) emitted from the two sources A (100 Hz frequency) and B (102 Hz frequency). Suppose that they are in phase at some arbitrary time ($t = 0$). At this time the resultant will be a maximum (Fig. 7.4a). When $t =$

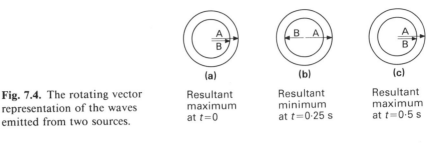

Fig. 7.4. The rotating vector representation of the waves emitted from two sources.

(a) Resultant maximum at $t=0$

(b) Resultant minimum at $t=0.25$ s

(c) Resultant maximum at $t=0.5$ s

0.25 s, A will have completed 25 vibrations and B 25.5; hence they will be out of phase and so a minimum will result. When $t = 0.5$ s, A will have completed 50 vibrations and B 51. At this time the sources will be in phase and the amplitudes will add once more (Fig. 7.4c). The period between beats is thus 0.5 s and the beat frequency is 2 Hz, i.e. the difference between the two original frequencies. *Note*: interference occurs here between the waves arising from two quite different sources. Two separate light sources on the other hand cannot produce interference patterns unless they are *coherent* (Section 8.10).

7.5 Sound production

1 *The human voice.* Sound is produced at the vocal chords which are stimulated to vibrate by air passing out through the larynx from the lungs. The fundamental note of a spoken word is set up by this process and the sound is then given quality when harmonics are set up in the resonating cavities of the pharynx, mouth and nose.

An amusing change in the quality of the spoken word can be brought about by inhaling helium before speaking. As the velocity of sound in helium is higher than that in air and as the characteristic wavelength produced by the cavities does not change, the *frequency of the resonance increases* in helium. The vocal chord frequencies on the other hand, being those of stretched membranes, do not change and therefore a normally deep voice will emerge as a high-pitched squeak.

2 *Insect acoustics.* Insects make great use of sound in communication and in the different species a wide range of frequencies is used.

The insect courtship sequence is often initiated by sound communication. The *Drosophila* male, for example, beats a wing in the direction of his chosen female and she tunes in, probably by using velocity sensitive hearing devices made up of sensory hairs which respond to displacements of the surrounding air. These receptors are located on the anal cerci and the antennae.

Insects make use of a very wide range of sound production and detection devices and Bennet-Clark's (1971) article is an excellent starting point for enthusiastic acoustic entomologists.

7.6   Sound receivers

There are basically two problems which any sound receiver, including the ear, has to overcome: (1) *reflection*, and (2) *transduction of sound energy into electrical or electrochemical energy*.

1   Sound waves travelling in air are very efficiently reflected when they impinge on a denser medium and this loss is a result of an impedance difference between the two media. For example, 99.9% of the sound energy is reflected at an air–water interface. The figure is the same for most biological materials, and it means that less than 0.1% of the incident sound energy is available for conversion into electrochemical energy. Modern microphones depend on very powerful electrical amplifiers to amplify the tiny voltages produced at a relatively mismatched transduction interface.

In the *ear*, however, there is some attempt to match up the impedance of the transducer to that of air. The first device is the ear flap or *auricle* which acts as a small ear trumpet. The sound energy is gathered from a large area and channelled into the smaller area of the *meatus* so that the forces available to set air molecules vibrating are much larger at the narrow end. This is equivalent to increasing the

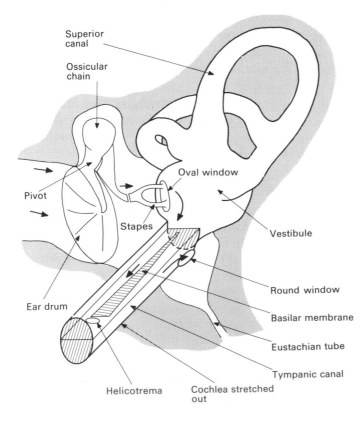

Fig. 7.5. Schematic diagram of the middle and inner ear in a mammal. The cochlea is shown uncoiled. The arrows show the displacements of fluid (air in the middle ear cavity and eustachian tube, perilymph in the vestibule and cochlea) produced by an inward movement of the ear drum. (After von Békésy, 1962).

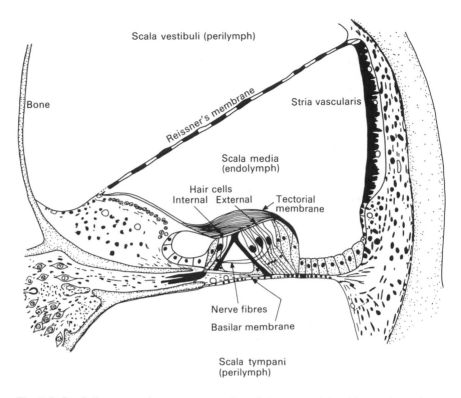

Scala vestibuli (perilymph)

Bone

Reissner's membrane

Stria vascularis

Scala media
(endolymph)

Hair cells
Internal External

Tectorial
membrane

Nerve fibres

Basilar membrane

Scala tympani
(perilymph)

Fig. 7.6. Semi-diagrammatic transverse section of the organ of Corti in a guinea pig (after Davis, 1953).

characteristic impedance of the narrow end, and means in fact that the impedance of the meatus is higher than that of the auricle (equation 7.4b). In this way the ear has begun to match up the impedances at the transducer interface. At the end of the meatus lies the ear drum (*tympanic membrane*) which is set vibrating by the sound waves. These vibrations are passed on through a system of levers called the *ossicular chain*, which is found in the middle ear. The *stapes* at the end of the chain beats against the oval window of the *cochlea* (Fig. 7.5). The mechanical advantage of the ossicles is approximately 2 so that the acoustic impedance of the system is once more increased because of the greater available forces and a further increment is achieved at the oval window, the area of which is only 1/20th that of the ear drum. The acoustic impedance from the air to the oval window has increased over 100 times by these devices. Waves are initiated in the cochlea canals by the vibration of the oval window; they are set up at one side of the cochlea in the perilymph of the scala vestibuli, travel through the helicotrema to the other channel, the scala tympani, and are dissipated at the round window (Figs 7.5 and 7.6).

2   The transduction mechanism whereby sound energy is converted into electrochemical energy at the *organ of Corti* (Fig. 7.6) is not

completely understood, but one plausible hypothesis makes use of the fact that waves will travel much faster in the cochlea fluid than they will in the relatively stiff basilar membrane. Hence when a pressure wave is set up it will move down the canal and through the helicotrema before the membrane has time to move. Because the pressures above and below the membrane will not be exactly in phase, there will be a pressure difference across the membrane which will displace it and cause a bulge. The actual position of the bulge will depend on the frequency of the sound. As the basilar membrane is thin and taut near the oval window and thick and slack towards the apex, high frequency sounds will cause the bulge to appear near the window, while low frequency sounds will cause a peak near the apex. The tectorial membrane moves up and down with the basilar membrane but there is a lateral shearing stress between them which will displace the hair cells. When this displacement occurs, the nerve fibres accompanying the hair cells generate action potentials and this acoustic information will be relayed through further fibres to the brain (von Békésy, 1960).

7.7 Echo-location

The ease with which bats fly and feed at night has intrigued scientists for over two centuries. It is now known that they emit strong high frequency sounds, ultrasonics, through their nostrils and detect the echoes from objects by means of their highly developed ears. Although bats have the best known echo-locating devices, the oil bird which lives in dark caves and the porpoise are only two examples of a wide range of species that possess this extra sense.

The precise mechanism involved in this extra sense is not completely understood but for the bats at least there are at present three hypotheses.

1 The simplest means of locating the distance of an object involves timing the interval between the emission of a high frequency signal and the arrival of the echo.

While cruising, bats of the *Vespertilionidae* family emit short high

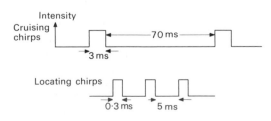

Fig. 7.7. Diagram of high frequency sound pulses produced by *Vespertilionidae* bats while echo-locating.

frequency sound pulses which are 3 ms long and about 70 ms apart (Fig. 7.7). As the velocity of sound in air is 3.3×10^2 m s^{-1}, if the echo arrived back 60 ms after emission this would tell the bat that the object was 10 m away.

As the bat approaches an obstacle, the repetition rate of the chirps increases to about 200 s^{-1}, and the chirp width decreases to 0.3 ms. This means that the echo from an object only 50 cm away will arrive just before the next emitted pulse. The fact that the sound frequency changes

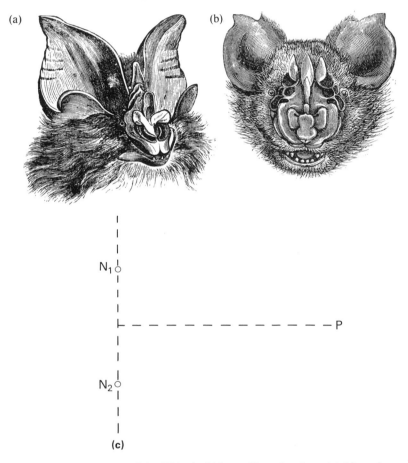

Fig. 7.8. Two examples of the *Rhinolophidae* or Horseshoe bats. (a) Mourning and (b) Trident bats from India and Iran respectively. These bats derive their name from the fact that their noses are formed from leaf-like membranous appendages, the front part of which is usually horseshoe-like in form. The nostrils are situated within this horseshoe, between it and other parts of the nose-leaf, which vary considerably in their shape and structure (from Duncan, 1880.) (c) Diagram explaining why the sound energy emitted from the nostrils (N_1 and N_2) of the Horseshoe bat is concentrated in front of the bat. N_1 and N_2 are spaced $\lambda/2$ apart. The sound energy at all points on a line drawn through N_1 N_2 is zero and hence the sound energy is concentrated in regions in front of the bat. The concave nostril flaps also help to focus the sound.

during the pulse from 110 kHz at the beginning to 40 kHz at the end probably means that some sophisticated frequency modulating system is also involved in the range finding of the *Vespertilionidae*.

2   The second probable method of echo-location depends on the ability of the bat to discriminate between echoes on an intensity basis. If the object is small compared to its distance from the bat, the intensity of the echo decreases as the fourth power of the distance of the bat from the object (Problem 7.2) and by simple successive measurements of the echo intensity, a bat could perceive whether it was approaching or receding from an object.

3   The third method, based on a change in the pitch of the echo, is probably used by the horseshoe bats, family *Rhinolophidae*. They emit very strong well-directed beams of sound of constant frequency, and the pitch of the returning echo will depend on whether the bat and the object are flying away from or towards one another. This frequency change is called the *Doppler effect* (see following section).

The nostrils of the horseshoe bats have two special features that allow them to emit the required narrow beams of high intensity. The first adaptation, and one which renders these bats startling in appearance, is a system of concave flaps round the nostrils which focus the sound waves in front of the bat Fig. 7.8 (a, b). The second feature is beautifully sophisticated and involves the spacing of the nostrils emitting the sound of constant frequency (and hence wavelength, λ). The bats have evolved nostrils that are spaced exactly $\frac{1}{2}\lambda$ apart and they therefore make use of the phenomenon of *wave interference* (Section 10). Consider any point P on the perpendicular which bisects N_1 and N_2, the nostrils of the bat (Fig. 7.8c), then if the sound waves start of at N_1 and N_2, in phase, they will arrive at P in phase and will give a maximum of intensity. It will also be seen that the intensity at all points on a line through N_1 and N_2 is zero as the path difference is always $\frac{1}{2}\lambda$. Hence most of the sound energy is concentrated directly in front of the nostrils.

7.8   The Doppler effect

In the days of steam, there was a well-known train-spotter's observation that the pitch of the train's whistle changed as it passed him in a station. As the train approaches, the observer hears a note which is higher than the true note and, on passing, the pitch quickly falls to a lower note than the true pitch. Doppler in 1842 was the first to give an explanation for this and there are three cases to consider.

1   *Stationary source and moving observer.* Let the observer be approaching the source with a velocity v_o (Fig. 7.9a). If the observer

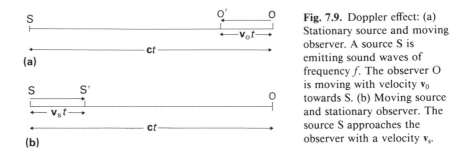

Fig. 7.9. Doppler effect: (a) Stationary source and moving observer. A source S is emitting sound waves of frequency f. The observer O is moving with velocity v_0 towards S. (b) Moving source and stationary observer. The source S approaches the observer with a velocity v_s.

were stationary at 0, f waves would pass him in 1 s. However, in 1 s he has advanced a distance v_0 to 0′, and he therefore hears in addition an extra vibration for each wavelength in the distance 00′, i.e. he hears $00'/\lambda$ or v_0/λ extra vibrations. Hence the apparent frequency is f' where

$$f' = f + \frac{v_0}{\lambda} = f + f\frac{v_0}{c} = f\left(1 + \frac{v_0}{c}\right). \tag{7.11}$$

If the observer had been receding from the source he would have heard fewer vibrations and the frequency would be given by

$$f' = f\left(1 - \frac{v_0}{c}\right). \tag{7.12}$$

2 *Moving source and stationary observer.* Let us assume that the source is moving towards the observer with velocity v_s. If f is the frequency of the source, then in a time t it will emit ft vibrations. At the end of t seconds suppose the first vibration has arrived at the observer, then the distance 0S will be equal to ct. The source will have advanced to S′ in this time and there will be ft vibrations between S′ and 0. With the source at rest there are ft vibrations between 0 and S and so the wavelength of the note which the observer hears from an approaching source is less than the wavelength for the source at rest. For an approaching source

$$\lambda' = S'0/ft = \frac{c - v_s}{f} \tag{7.13}$$

but from equation (7.1)

$$\lambda' = c/f'$$

hence $\qquad \dfrac{c - v_s}{f} = c/f' \tag{7.14}$

or $\qquad f' = f\left(\dfrac{c}{c - v_s}\right).$

If the source is receding from the observer we obtain

$$f' = f \left(\frac{c}{c + v_s} \right). \tag{7.15}$$

3 *Both source and observer moving.* If the observer (v_o) and the source (v_s) are both approaching a fixed point, then the apparent pitch is

$$f' = f \frac{c + v_o}{c - v_s}. \tag{7.16}$$

The algebraic sign of either numerator or denominator is changed if the direction of either v_o or v_s is reversed.

Problem 7.1

(a) When sound waves meet a boundary between two media, the ratio of the transmitted to incident intensities is given by the equation (Alexander, 1968)

$$I_t / I_i = 4 \rho_1 c_1 \rho_2 c_2 / (\rho_1 c_1 + \rho_2 c_2)^2. \tag{7.17}$$

Show that for an air–water interface only 0.1% of the incident energy is transmitted.

$$\rho_{air} = 1.3 \text{ kg m}^{-3} \qquad\qquad \rho_{water} = 10^3 \text{ kg m}^{-3}$$

$$c_{air} = 3.3 \times 10^2 \text{ m s}^{-1} \qquad\qquad c_{water} = 15 \times 10^2 \text{ m s}^{-1}.$$

(b) If the acoustic impedance of the air is increased 100-fold by various devices, how much incident energy is now transmitted through the interface?

Problem 7.2

(a) The inverse square law states that if E is the energy emitted by a point source, then the energy falling on unit area of a surface some distance r from the source is inversely proportional to r^2. Show that the intensity of an echo from a point source, at the source, is inversely proportional to r^4.

(b) Consider two stationary objects which reflect echoes of the same intensity when a bat is 0.3 m from the smaller and 1 m from the larger. Show that after the bat has flown 5×10^{-2} m towards them, the echo intensity from the smaller will be doubled, whereas that from the larger will be increased by only 20%.

Problem 7.3

(a)   A flying bat is chasing a moving object. The bat is producing high intensity sounds, listening to the echoes, and is measuring the time between the echoes as it approaches the object. If α is the fractional increase in echo intensity from one echo to the next and τ seconds is the time between echoes, show that the bat has to fly for t seconds after the arrival of the second echo before catching up with the object, where t is given by

$$t = \frac{\tau}{(1 + \alpha)^{1/4} - 1} \text{ s.} \tag{7.18}$$

You may consider that the bat and object are flying with constant velocities.

(b)   If there is a 20% increase in echo intensity with 250 ms between echoes, show that the bat has to travel for 5 s before it reaches the object.

Problem 7.4

A bat flies straight towards a wall at a speed of 10 m s^{-1} while emitting a steady note of frequency 42×10^3 Hz. What frequencies does it hear? ($\mathbf{c} = 3.3 \times 10^2 \text{ m s}^{-1}$).

7.9   Ultrasound

Ultrasound is used in medical diagnosis as there are no known harmful effects on tissues at the power levels normally employed. The principles of ultrasonic diagnosis are exactly the same as those of echo-location in bats. A short pulse of ultrasound is emitted from a transducer and after a short time delay an echo is obtained at the receiver. The transducer is commonly made from piezo-electric materials (Fig. 7.10), such as

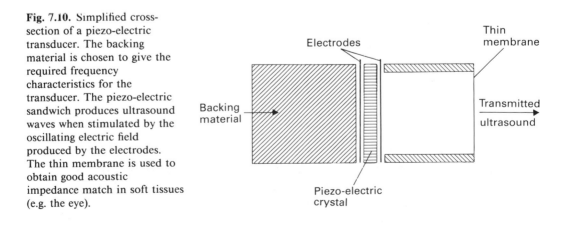

Fig. 7.10. Simplified cross-section of a piezo-electric transducer. The backing material is chosen to give the required frequency characteristics for the transducer. The piezo-electric sandwich produces ultrasound waves when stimulated by the oscillating electric field produced by the electrodes. The thin membrane is used to obtain good acoustic impedance match in soft tissues (e.g. the eye).

Electrodes

Thin membrane

Backing material

Transmitted ultrasound

Piezo-electric crystal

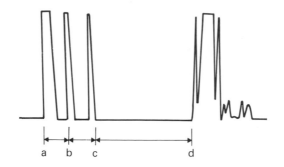

Fig. 7.11. Ultrasound A-scan echogram of the human eye. The major echoes from the cornea (a), lens (b and c) and sclera and retina (d) can clearly be seen.

quartz, tourmaline and, more commonly, lead titanium zirconate, which change shape when subjected to an electric field. The deformations induced by high frequency electric fields set up high frequency sound waves. As piezo-electric materials also produce electric fields when they are deformed, they can also be used to form the receiver and for most purposes in medical ultrasound the same crystal is used for both transducer and receiver.

Because air and tissues have very different acoustic impedances, ultrasound energy losses are minimized either by encasing the transducer in a water tube before placing it on the tissue or organ to be examined, or, in the case of abdominal scanning, for example, by smearing the skin with castor oil or vaseline before applying the transducer directly.

Ultrasound machines generate basically two types of display — the A and B scans respectively. The former gives a display of echo amplitude versus time directly, while the B scan gives a two-dimensional image of the object either by moving the transducer around the object of interest or by using a linear array of a number of transducers. Each element is a transducer and they can be fired in turn to build up a complex two-dimensional image.

The ultrasonic echogram (A-scan) of the eye, for example, shows very clearly the echo associated with the cornea (a) anterior and posterior faces of the lens (b and c) and the retina choroid complex at the back of the eye (Fig. 7.11). When compared with the quantitative slit lamp camera photograph (Fig. 8.22) it can be seen that surfaces which scatter light efficiently also scatter ultrasound.

7.10 The B-scan (brightness modulated scan)

In a B-scan, the amplitudes of the reflected signals are displayed on a cathode ray screen (or video monitor) as spots of variable brightness, with larger signals producing brighter spots. If a single transducer element is used then it is moved continuously in a plane around the

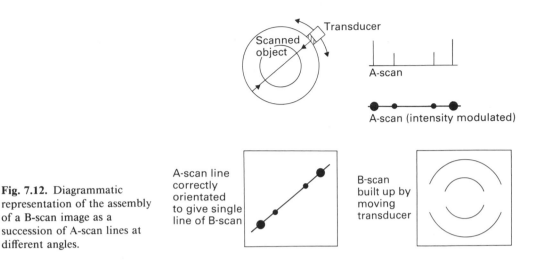

Fig. 7.12. Diagrammatic representation of the assembly of a B-scan image as a succession of A-scan lines at different angles.

patient. This movement generates a two-dimensional image of the patient provided that the position of the probe in space is accurately known and related to a position on the display screen. In this way the reflection from a surface at one position of the transducer will register in exactly the same position on the screen irrespective of the position from which it is detected (Fig. 7.12).

Real-time scanning

In real-time scanning it is possible to study the movements of organs and vessels by using multi-element transducers. In this way a two-dimensional B-scan is rapidly built up without lateral movement of the transducer head.

Fetal scanning is probably the most widely used application of ultrasound in medicine and, in the more sophisticated machines, computers are employed for image reconstruction. Using a multi-element transducer coupled to a computer imaging system it is possible, for example, to 'see' a fetus yawn in the womb (Fig. 7.13).

Doppler ultrasound (angiodynography)

In Section 7.8 we have seen that the perceived sound frequency depends on the relative motions of the source and the observer. This change in frequency can be used, for example, to detect red cells moving in a blood vessel. If an ultrasound transducer is emitting a signal of frequency f_t, then it will record a frequency f_r scattered back from red cells moving with velocity **v** (Fig 7.14). The Doppler shift in frequency $(f_r - f_t)$ is given by

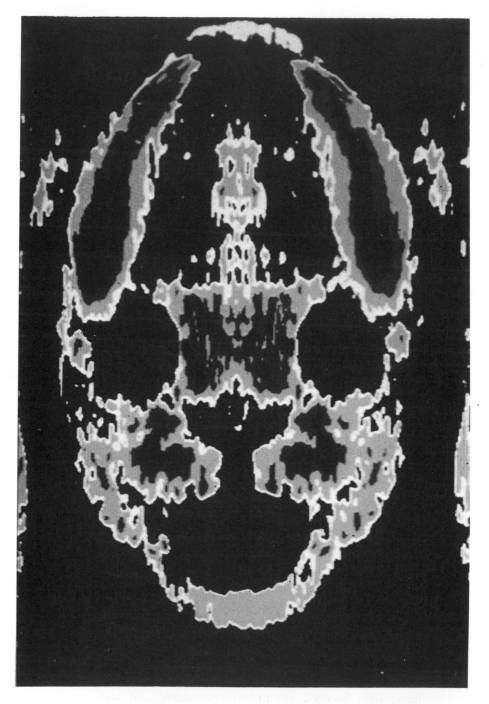

Fig. 7.13. Ultrasound image of a fetal head in the sixth month of pregnancy. It is possible to see that the mouth is wide open. (Reproduced from *The National Geographic* by kind permission of the publishers and Professor C.R.B. Merritt.)

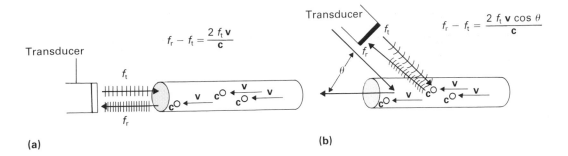

(a)   **(b)**

Fig. 7.14. Doppler effect describes the change in frequency of ultrasound scattered by red cells moving within a vessel. (a) If the ultrasound signal is transmitted along the axis of the vessel, then the velocity of the red cells (v) can be related to the shift in Doppler frequency $f_r - f_t$, where f_r is the frequency of the reflected signal, f_t is the frequency of the transmitted signal, and c is the velocity of sound in the medium. (b) If the signal approaches the vessel at an angle θ, then the frequency of the reflected signal f_r is reduced in proportion to the cosine of θ.

$$f_r - f_t = \frac{2f_t \mathbf{v}}{c},$$

when the red cells are moving directly towards the transducer receiver

$$\text{while } f_r - f_t = \frac{2f_t \mathbf{v} \cos \theta}{c}$$

if the cells are moving at an angle θ with respect to the transducer (Fig. 7.14). Examples of the use of conventional Doppler ultrasound in the examination of blood flow characteristics in the normal and diseased carotid artery have already been given in Chapter 5 (Fig. 5.12)

An exciting innovation in this field has been the combining of B-scan imaging with Doppler analysis to give very beautiful and detailed real-time images of blood flow patterns.

In most *angiodynography* systems, high frequency ultrasound pulses are produced by a linear array transducer. The amplitude information in the back-scattered sound waves is used to form the B-mode image. Rapidly moving targets, such as red cells, produce echoes of such low amplitude that they are not commonly displayed. However, since a change in frequency occurs with moving objects, an analysis of the phase and frequency of the echoes will give a detailed map of the velocity profile through the lumen of the vessel.

Interestingly, there are two conflicting requirements in transducer design for angiodynography. The most detailed B-scan images are obtained when the transducer is at right angles to the vessel wall, while the greatest Doppler shift is obtained when the axis of the transducer

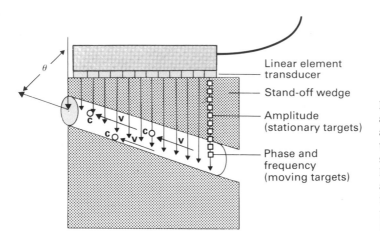

Linear element transducer

Stand-off wedge

Amplitude (stationary targets)

Phase and frequency (moving targets)

Fig. 7.15 Diagram of angiodynography transducer arrangement. The linear array transducer produces 'line of sight' ultrasound pulses. Each line is sampled for echo signal amplitude, phase and frequency. The stand-off wedge provides the Doppler angle to image blood flow.

and the direction of flow are parallel. Because many vessels lie parallel to the skin surface, compromises are required to combine the two images. This is usually provided by a stand-off wedge of material between the transducer array and skin surface (Fig. 7.15). This defines the Doppler angle (θ) for vessels lying parallel to the surface.

Fig. 5.13 gives an example of the use of angiodynography methods in mapping blood flow profiles in the diseased carotid artery. Modern machines display colour-coded Doppler images of the blood vessels together with graphs of the computed flow velocity at specific points in the vessel, throughout several cardiac cycles. The display therefore consists of a combination of Figs 5.12 and 5.13. Numerous examples of this powerful technique are given in Merritt (1986 and 1987) and in Powis (1986).

References

Alexander, R.M. (1968) *Animal Mechanics.* Sedgwick & Jackson, London.

Bennet-Clark, H.C. (1971) Acoustics of insect song. *Nature*, **234**, 255–9.

Davis H. (1953) Acoustic trauma in the guinea pig. *Journal of the Acoustics Society of America*, **25**, 1180–89.

Duncan, P.M. (1880). *Natural History*, Vol. 1, pp. 259–341. Cassell, Petter & Galpin, London.

Merritt, C.R.B. (1986) Doppler blood flow imaging: integrating flow with tissue data. *Diagnostic Imaging*, **8**, 146–55.

Merritt, C.R.B. (1987) Doppler colour flow imaging. *Nature,* **328**, 743–4.

Powis, R.L. Angiodynography. A new real-time look at the vascular system. *Applied Radiology*, Jan/Feb. 1986.

von Békésy, G. (1962) The gap between the hearing of internal and external sources. *Symposium of the Society for Experimental Biology,* **16,** 267–88.

von Békésy, G. (1960) *Experiments in Hearing.* McGraw-Hill, New York.

Further reading

Ackerman, E. (1962) *Biophysical Science.* Prentice-Hall, London.

Aidley, D.J. (1989) *The Physiology of Excitable Cells*, 3rd edn. Cambridge University Press.

Aidley, D.J. (1969) Echo intensity in range estimation by bats. *Nature*, **224**, 1330–1.

Brown, B.H. & Smallwood, R.H. (1981). *Medical Physics and Physiological Measurement.* Blackwell Scientific Publications, Oxford.

Dendy, P.P. & Heaton, B. (1987) *Physics for Radiologists.* Blackwell Scientific Publications, Oxford.

Sears, F.W. & Zemansky, M.W. (1964) *University Physics.* Addison-Wesley, Reading, Mass.

Sochurek, H. (1987) Medicine's new vision. *National Geographic,* **171**, 2–41.

8 Optics and Microscopy

8.1 Reflection and refraction

Historically, the first step in the scientific study of light was made by Euclid in 300 BC who wrote: 'Light travels in straight lines called rays.' On this is based the science of geometrical optics and briefly, the two most important laws are:

1 *The Law of Reflection* states that when a ray of light is reflected from a plane surface, the angle of incidence equals the angle of reflection (Fig. 8.1).

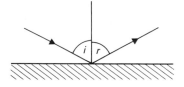

Fig. 8.1. Reflection at a plane surface. The angle of incidence i equals the angle of reflection r.

2 *The Law of Refraction* states that when light rays pass from one medium to another, the angle of refraction (r) is not equal to the angle of incidence (i). In fact Snell discovered that the sine of the angle of incidence bears a constant ratio to the sine of the angle of refraction. The ratio is known as the *refractive index* from one medium (1) to another (2) and is denoted by $_1\mu_2$, i.e.

$$\frac{\sin i}{\sin r} = {}_1\mu_2. \tag{8.1}$$

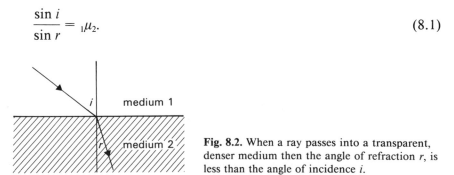

Fig. 8.2. When a ray passes into a transparent, denser medium then the angle of refraction r, is less than the angle of incidence i.

Usually the refractive index of a material is expressed relative to a vaccum, or air, which is taken as 1. In fact light is slowed down in passing from air to a medium of higher refractive index μ; if **c** is the velocity in air, then **c**/μ is the velocity in the medium. Newton was

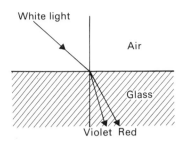

Fig. 8.3. On refraction, white light is split into its constituent colours. The violet end of the spectrum is deviated most and the red is deviated least.

among the first to show that different wavelengths of light are refracted by different amounts and at an air–glass interface violet is deviated most and red least (Fig. 8.3).

8.2 Total internal reflection

When a light ray passes from a dense to a less dense medium both refraction and total internal reflection can occur (Fig. 8.4). In this case the refractive index ($_1\mu_2$) is less than unity and let us consider what happens when rays originating from O, strike the interface at increasing angles. Since the angle of refraction (r_1) must be greater than the angle of incidence (i_1), a limiting condition is reached when the angle of refraction (r_c) is 90° and the ray lies parallel to the surface. i_c is known as the *critical angle* and provides a convenient way of measuring the refractive index since:

$$\sin i_c = {}_1\mu_2.$$

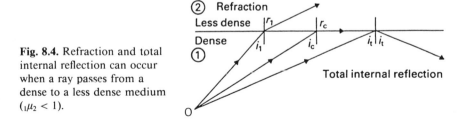

Fig. 8.4. Refraction and total internal reflection can occur when a ray passes from a dense to a less dense medium ($_1\mu_2 < 1$).

For all angles greater than i_c, total internal reflection occurs.

The principle of total internal reflection is used in the manufacture of prism binoculars, periscopes and fibre optics. The latter are manufactured from long lengths of glass or plastic, cemented together with a low refractive index resin or cladding. If near parallel light rays are focused on one end of a fibre, then they will be repeatedly reflected at the interface until they reach the other end (Fig. 8.5). Optic fibres can be made of a pliant material and the individual fibres can be cemented into a coherent pattern. This means that the position of any one fibre

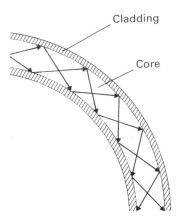

Cladding

Core

Fig. 8.5. The transmission of light rays along an optic fibre. The core and cladding materials have different refractive indices to allow efficient internal reflection. Different combinations are possible. For example, a silica core with plastic cladding provides a rugged, flexible fibre, while fibres with a glass core (high density) and glass cladding (low density) are used for illumination and picture transmission applications.

relative to the others is maintained at either end of the bundle and so images can be transmitted along the bundle. Optical fibres are in widespread use in medicine as they can be used to view the internal tracts of the human body.

8.3   The rainbow

The basic requirements for a rainbow to be seen are that it should, of course, be raining, but the sun must be behind the observer. The

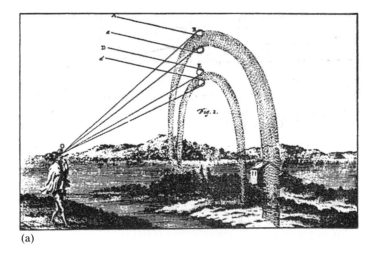

(a)

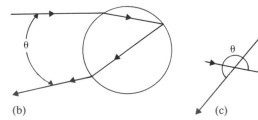

(b)   (c)

Fig. 8.6. (a) Studying the geometry of a double rainbow: a scientific engraving from the mid-18th century. (b) Ray path through a raindrop in the primary (lower) rainbow. (c) Double internal reflection path in the secondary bow.

scientific engraving shows this geometry and also traces the ray paths of both the primary and secondary rainbows. In the primary (Fig. 8.6b), the ray is first refracted at the air–water interface, by an amount depending on its wavelength, then totally internally reflected before being again refracted towards the observer. The angle of deviation $(180° - \theta)$ depends on the wavelength of light and it varies from 138° for red light $(\theta = 42°)$ to 140° for violet $(\theta = 40°)$. This explains why the colour is red on the outside of the rainbow and violet on the inside.

The secondary bow is produced by two internal reflections (Fig. 8.6c) and in this case the angle of deviation *is* the angle θ. The corresponding angles are 50.5° for red and 54° for violet, so in the secondary bow the colours are reversed with red on the inside and violet on the outside.

The perfect symmetry of the rainbow in fact shows that each raindrop is perfectly spherical and this shape arises from surface tension forces at the water–air interface (Section 6.1)

8.4 Mirrors and lenses

Curved reflecting and refracting surfaces can focus light to form images. In the case of a mirror whose surface is a part of the surface of a sphere, the equation relating the distance of the object from the mirror, u, the image distance v, the radius of curvature of the mirror r, and the focal length of the mirror f is

$$\frac{1}{v} + \frac{1}{u} = \frac{1}{f} = \frac{2}{r}.$$
(8.2)

There are several sign conventions in use and one quite good one is that distances are measured from the mirror and they are regarded as positive if they are in the same direction as that of the incident light and negative if they are in the opposite direction. A *concave* mirror focuses parallel light on to the focal point in front of it (Fig. 8.7a) while a *convex* mirror causes the rays to diverge so that they appear to come from a

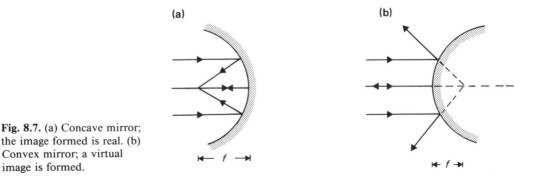

(a)   **(b)**

Fig. 8.7. (a) Concave mirror; the image formed is real. (b) Convex mirror; a virtual image is formed.

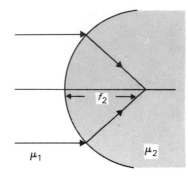

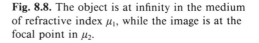

Fig. 8.8. The object is at infinity in the medium of refractive index μ_1, while the image is at the focal point in μ_2.

point behind it (Fig. 8.7b). Using the above sign convention, the focal length of a concave mirror is negative, while that of a convex mirror is positive. The magnification of a mirror system is given by the relationship

$$\text{image size/object size} = v/u. \tag{8.3}$$

Curved refracting media also focus light (Fig. 8.8). As the cornea and aqueous humour have similar refractive indices, the lensless eye is an example of such a system. The equation to apply in this case is

$$\frac{\mu_2}{v} - \frac{\mu_1}{u} = \frac{\mu_2 - \mu_1}{r} \tag{8.4}$$

With the sign convention as before, μ_1 and μ_2 are the refractive indices of medium 1 and medium 2 with respect to air and r is the radius of curvature of the interface.

The magnification of such a system is

$$\frac{\text{image size}}{\text{object size}} = \frac{v\mu_1}{u\mu_2} \tag{8.5}$$

Lenses (Fig. 8.9) consist of a homogeneous refracting medium bounded by spherical surfaces. The equation for thin lenses is

$$\frac{1}{v} - \frac{1}{u} = \frac{1}{f} \tag{8.6}$$

The *power* of a lens is defined as the reciprocal of the focal length and the units are *dioptres* with units m^{-1}. The lens of the eye approximates only very crudely to a thin lens, as not only is it quite thick, but its refractive index is non-homogeneous and increases towards the centre of the lens.

The magnification of a thin lens is given by

$$\text{image size/object size} = v/u. \tag{8.7}$$

There are two types of lens (Fig. 8.9), namely converging and diverging,

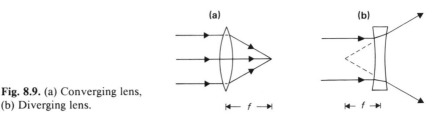

Fig. 8.9. (a) Converging lens, (b) Diverging lens.

and they have positive and negative focal lengths respectively. In order to solve problems involving lenses, the same sign convention as that used for mirrors can be adopted.

8.5 The eye as an optical system

The eye (Fig. 8.10) can be compared with a camera. It has a refracting system consisting of the cornea and lens which forms inverted images of viewed objects on the back surface of the retina. Except for the transparent cornea, a tough opaque coating called the sclera covers the eye and inside this is the dark pigmented layer, the choroid, which prevents stray light from being scattered around the eye. There is a blind spot in the eye where the optic nerve leaves.

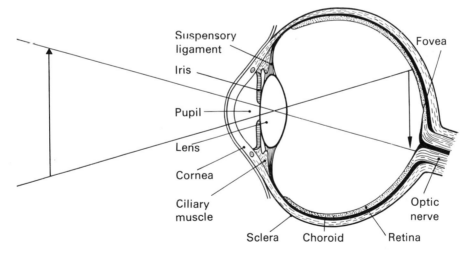

Fig. 8.10. Diagrammatic cross-section of the human eye. An inverted image is formed on the retina at the back of the eye. In terrestial animals the main focusing device is the cornea because the main refractive index difference occurs at the cornea–air interface. In amphibian and aquatic species, however, the lens plays a greater role and, in these species, it is usually harder and more spherical. Accommodation in these animals is usually achieved by moving the lens nearer, or further away from, the retina rather than by altering the shape of the lens. In all vertebrate types, however, it always seems somewhat perverse that the light has to pass through several nerve cell layers before reaching the photoreceptors cells, especially as several lower animals with a similar eye structure (e.g. squid and octopus) have managed to evolve a retinal system where the photoreceptors are at the front surface of the retina.

In terrestrial animals the main refraction occurs at the cornea while the lens provides a fine adjusting device. In man the focusing power provided by the cornea is approximately 42 dioptres while in young adults the lens power is variable from about 23 to 33 dioptres. The ciliary muscles control the shape of the lens and hence *the accommodation of the eye.* The focal length of the lens is limited partly by the elasticity of the lens and partly by the strength of the ciliary mucles. Since the focal length of the lens cannot be decreased indefinitely, the eye cannot focus on the retina images of objects closer than a certain distance, the *near point* of the eye, and for a young adult this is approximately 0.25 m. The furthest point that can be focused on the retina is termed the *far point.*

If D and d are the distances in metres from the eye of the far and near points respectively, then $1/d - 1/D$ is called the *amplitude of accomodation* of the eye and is measured in dioptres. This decreases with age as a result of a loss in elasticity of the lens. For a ten-year-old child, $D = \infty$ and $d = 7 \times 10^{-2}$ m and so the amplitude of accomodation of the eye is 14 dioptres.

In a camera, light is focused on to light-sensitive silver bromide emulsion film. In the eye, the light-sensitive surface takes the form of photopigments located in basically two types of photoreceptor cells, namely *rods* and *cones.* The receptor cells record the arrival of light energy at the retina; this information is processed to some extent in retinal neurones, and the information is carried for further processing to the brain by ganglion cells which leave the eye in a bundle at the blind spot. Many rod receptors feed one ganglion cell so there is a pooling of visual information with a consequent increase in sensitivity. However this pooling leads to a loss in *visual acuity,* i.e. in apparent sharpness and detail of the object that is being viewed. Rod cells contain only one photopigment, called *rhodopsin* and so no colour vision is possible using these receptors. On the other hand, there are three types of cone receptors: blue, green and red/yellow, each containing a different pigment so that colour vision is possible with this system. The cones are concentrated around the optic axis of the eye at the fovea and these are the receptors that are mainly used when looking straight at an object. As there is more or less one ganglion cell for each cone, visual acuity is high for the cones, but the level of sensitivity is much less than for the rods.

Apart from having the ability to switch from one type of receptor to another, the eye has further advantages as it is linked to sophisticated physiological data processing and storing devices in the retina and brain. This means that we can relate the image presented on our retina with past visual experiences. Also if we are presented with conflicting visual evidence we continuously scan the picture trying to make sense of it. Hence the image of Fig. 8.11 which we see in our mind continually

Fig. 8.11. When two conflicting images are presented to the eye we see alternatively one or the other — in this case an old or young woman.

oscillates between the two possibilities in the picture, namely that of either a young or an old woman.

8.6 Visual acuity

Visual acuity is a measure of the ability of the visual system to resolve a target and this resolution is dependent on three main factors: the background illumination, the target illumination and the visual angle the target subtends at the nodal point of the eye. It is important, therefore, that during routine measurements of visual acuity all these parameters are carefully controlled. The first two involve careful

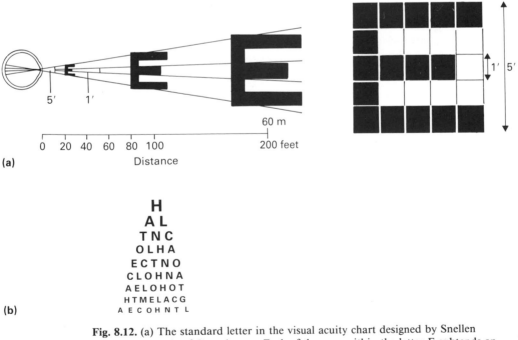

Fig. 8.12. (a) The standard letter in the visual acuity chart designed by Snellen subtends an angle of 5′ at the eye. Each of the gaps within the letter E subtends an angle of 1′. (b) Letters are mounted on the standard chart in order of decreasing size.

attention to light conditions during the examination and the third involves the construction of a special chart consisting of groups of letters of different sizes (Fig. 8.12). The letters are constructed so that they will subtend the same visual angle when viewed at distances of up to 60 m (200 ft). The letters are normally mounted on card or in a system that allows them to be illuminated from behind (Snellen Charts). The top row of letters can be read at a distance of 60 m by a person with normal visual acuity, while the bottom row can only normally be read at a distance of 6 m (20 ft). The patient is seated 6 m from the chart and his visual acuity is determined from the last row of letters that he can see clearly. The visual acuity is usually expressed as a ratio. For example if his vision is so poor that he can only see the top line clearly, his VA score is 6/60. This score is the ratio of his distance from the chart divided by the greatest distance from which a person with normal sight would be able to see the letter. If he had normal vision, the VA score would be 6/6 (or 20/20 in North America).

The measured visual acuity depends on the position of the image on the retina as the density of the rods and cones varies with distance from the fovea (Fig. 8.13). The cones are highly concentrated in the region $\pm 10°$ from the centre of the fovea while the rods are almost entirely absent from this region. Both, of course, are absent where the optic nerve exits through the retina (18° from the fovea). The rods probably represent the form of vertebrate photoreceptor system that evolved first and although they provide relatively poor visual acuity, they and their attendant processing neurones (horizontal, bipolar, amacrine and ganglion cells) provide an excellent movement detector system. We are therefore well equipped to detect potentially predatory movement 'out of the corner of an eye'.

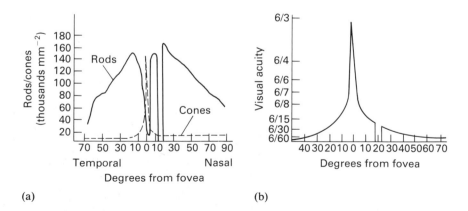

(a)   (b)

Fig. 8.13. (a) Rod and cone density plotted as a function of distance from fovea. (b) Variation of visual acuity with distance from fovea.

8.7 Lens defects and loss of visual acuity

Although it represents an extremely sophisticated receptor system, the eye may suffer from certain defects which result in a loss of visual acuity. The first two of these, namely *spherical* and *chromatic* aberration, are common to all homogenous lens systems and in the eye there are inherent correction devices which make the eye surprisingly free from these defects. The other visual defects which include short and long sight, presbyopia, astigmatism and cataract are corrected by man-made devices and interventions.

1 *Spherical aberration*

This is a property of all lenses bound by spherical surfaces. The marginal portions of the lens bring rays to a shorter focus than the central region (Fig. 8.14). The image of a point is therefore not a point but a small blurred circle.

Fish have probably the densest lenses and so would be expected to suffer commensurately from spherical aberration. The fact that they do not (Fig. 8.15) was appreciated by early 19th century physicists and Sir David Brewster went so far as to suggest that a very superior microscope

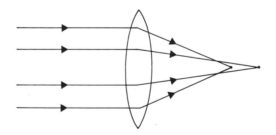

Fig. 8.14. Spherical aberration. Light from the margins is brought to a shorter focus than that from the central regions.

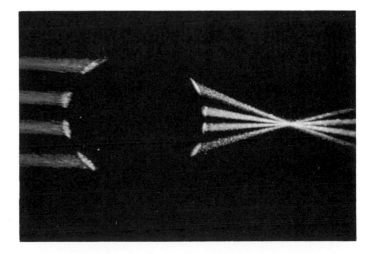

Fig. 8.15. Four laser beams focused by the crystalline lens of the African cichlid fish *Haplochromis burtoni*. (From Fernald & Wright, 1983.)

could be constructed using fish lenses. Brewster and others suggested that the lack of spherical aberration was due to a gradient of refractive index from the centre (highest) to the outer surface (lowest value) of the lens. Recent elegant experiments employing collimated laser beams have confirmed the remarkable lack of spherical aberration in the teleost lens (Fig. 8.15).

The human eye in fact corrects for spherical aberration by means of two mechanisms that have evolved with the eye:
(a) As in fish (and other aquatic animals) the lens is denser in the centre and hence refracts light more strongly at its core than at its outer layers.
(b) In terrestrial animals, where the cornea plays a relatively more important role, the cornea is flatter at its margin than at its centre.

2 Chromatic aberration

All lenses made of a single, homogeneous material refract rays of shorter wavelength more strongly than those of longer wavelength, and so bring blue light to a shorter focus than red (Fig. 8.16). The result is that the image of a point of white light is not a white point, but a blur circle fringed with colour. Since this seriously disturbs the image, even the lenses of inexpensive cameras are corrected for chromatic aberration. The error is actually moderate between the red end of the spectrum and the blue/green, but it increases rapidly at shorter wavelengths — the blue, violet, and ultraviolet (Fig. 8.17).

As in the case of spherical aberration the eye has overcome these inherent lens defects by evolutionary processes. The first device helping the eye is the lens which acts as a colour filter. It passes the visible spectrum, but cuts off sharply at the violet end, where chromatic aberration is worst. The remaining devices are to be found in the retina itself. In 1825 Purkinje noticed that at the first light of dawn, blue objects tend to look relatively bright compared with red objects, but then they tend to look relatively dim as the morning advances. The basis of this change is in the difference in spectral sensitivity between the two types of photoreceptor cells (Fig. 8.17). Rods are maximally sensitive in the blue/green, i.e. 500 nm, whereas cones have their maximum in the green

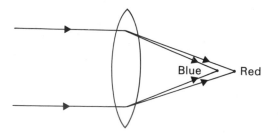

Fig. 8.16. Chromatic aberration. Shorter wavelengths are brought to a closer focal point than long wavelengths.

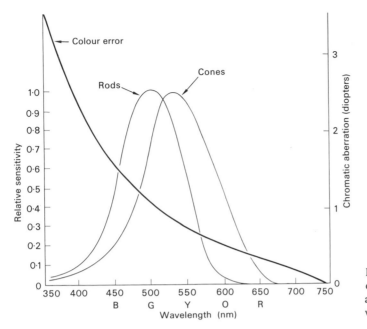

at 540 nm. Hence as one goes from dim light to bright light where the pattern vision is good the sensitivity of the eye moves away from the region of the spectrum in which the chromatic aberration is large.

The third device is to be found in the fovea where the concentration of cones is highest. In man, apes and monkeys alone of all known mammals, the fovea and the region round it, called the macula lutea, is coloured yellow. The yellow pigment is xanthopyll, a carotenoid that also occurs in all green leaves. The pigment absorbs maximally in the violet and blue regions of the spectrum, just where absorption by the lens falls to low values, and hence removes further parts of the spectrum where chromatic aberration is worst.

3 Short-sight (myopia)

A short-sighted person can see near objects distinctly, but not distant objects. The latter are focused in front of the retina as the eyeball is too long (Fig. 8.18a). If it is supposed that the far point is 1 m from the eye, application of equation (8.6) gives the type of spectacles to be prescribed in order that objects at infinity may be clearly seen,

$$\frac{1}{f} = \frac{1}{v} - \frac{1}{u}.$$

The lens must focus the parallel rays of light at the far point in order

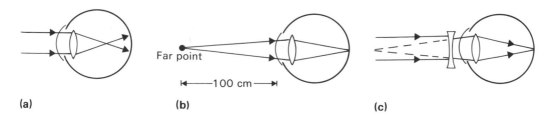

(a)   **(b)**   **(c)**

Fig. 8.18. The short-sighted eye. (a) In a short-sighted person, the eye is too long and so the light from a distant object (parallel lines) is brought to a focus in front of the retina. (b) The furthest distance from the eye that an object can be brought to focus on the retina is called the far point. (c) In order to focus parallel light on the retina a diverging lens is used which forms a virtual image at the far point of the eye. The cornea–lens system of the eye can now focus light on the retina.

that the cornea/lens system can focus it on the retina. When $v = -1$ m, $u = -\infty$, then $f = -1$m.

Hence a concave diverging lens of 1 dioptre power must be prescribed.

4 *Long-sight (hypermetropia)*

A long-sighted person can see distant objects distinctly but not near objects; the latter are focused behind the retina as the eyeball is too short (Fig. 8.19a). If it is supposed that the near point is about 0.6 m from the eye, it is left as an exercise to show that converging spectacles are required to form a virtual image 0.6 m distant of an object at a comfortable viewing distance 0.25 m from the eye (Fig. 8.19c).

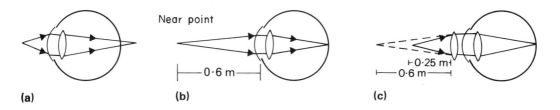

(a)   **(b)**   **(c)**

Fig. 8.19. The long-sighted eye. (a) In this case the eyeball is too short and light from a near object is focused behind the eye. (b) An object at the near point can, however, be clearly seen. (c) Converging spectacles are used when viewing closer objects and these form a virtual image at the near point.

5 *Far-sight (presbyopia)*

As the eye ages, the lens loses some of its elasticity so that the amplitude of accomodation is reduced in many cases to only 1 dioptre at 60 years of age. The lens shape cannot be changed to focus on near objects and, to compensate, converging spectacles are required for reading. However, a short-sighted presbyope requires bifocal spectacles; the upper half of each lens is diverging for long distance vision, while the lower half corrects for presbyopia.

6 *Astigmatism*

This occurs as a result of a lack of symmetry in the cornea and if a pattern such as that in Fig. 8.20 is viewed, one set of lines will appear sharper than the others. This defect can be corrected by cylindrical lenses.

Fig. 8.20. Astigmatism. To a person viewing this pattern, and suffering from astigmatism, one set of lines will appear sharper than the others.

7 *Cataract of the lens*

The most common cause of loss of visual acuity throughout the world is cataract of the lens. It is estimated that 35 million people are blind through this disease and at present the only cure is through surgical removal of the lens.

Since the cornea forms a transparent window into the eye, cataract development can be followed with the aid of a slit lamp camera. A narrow beam of light is projected on to the lens and the scattered light can be viewed directly at an angle to the eye or recorded photographically. Note that a tilted film plane camera has to be used if the entire optical section of the lens is to be in focus on the film plane (Fig. 8.21). The two major forms of cataract are nuclear and cortical. The first involves a change in the proteins of the central regions of the lens. The modified proteins both scatter and fluoresce in the visible region of the spectrum, so nuclear cataracts appear coloured. The second type involves changes in the outer regions of the lens and since only light-scatter is involved, these regions appear white. The increase in scatter appears to be due partly to an aggregation of proteins to form large light scattering particles and partly to an increase in lens water content causing abrupt refractive index changes within the outer regions (Fig. 8.22).

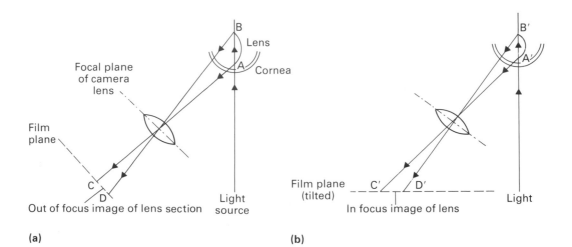

(a)   **(b)**

Fig. 8.21. Diagram to show optics of tilted film plane slit lamp camera. A narrow beam of light is directed into the eye through the centre of the cornea. Light rays reflected from the two corneal interfaces and the front and back surfaces of the lens (A and B) are focused by a camera lens on to the film (a) in a conventional camera and (b) in a camera where the back of the camera has been adapted so that the film plane is perpendicular to the rays from the light source. In (a) the light path AC is shorter than BD, so the whole of the lens section cannot be imaged at one time. In (b) the paths are identical, so the lens (and corneal) reflections can be focused in one plane.

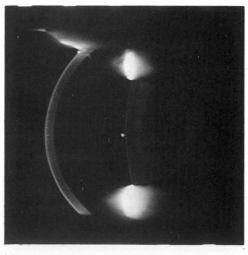

(a)

Fig. 8.22. (a) Slit-lamp camera photograph of a normal eye. Note the lack of light scatter from the lens. The air/cornea interface scatters light because of the major refractive index change (*continued opposite*).

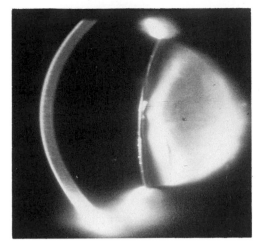

(b)

(c)

Fig. 8.22 (*continued*) Slit-lamp camera photographs of (b) cortical, and (c) nuclear cataracts (from Duncan, 1989).

8.8 Comparative physiology of the visual system

It is well known that nocturnal animals have a largely all-rod retina (e.g. rat), while diurnal animals have a high density of cones, especially in the regions centred round the optic axis (e.g. man, pigeon). However, subtle differences also occur at the lens level. In diurnal animals such as man, the fetal lens is near spherical and as adulthood approaches the lens gains a more elliptical form (Fig. 8.23). In this way the power of the lens decreases in line with the growth of axial length of the eye (Fig. 8.24a). However, in the rat, although the fetal eye is similar in shape to that of the fetal human, as the eye ages the lens becomes increasingly spherical and the lens to retinal distance decreases (Fig. 8.25).

In general, the lens of the adult terrestrial vertebrate is elliptical and also soft to allow for accommodation. The typical fish lens is a never-

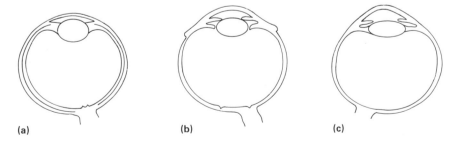

(a)   (b)   (c)

Fig. 8.23. Drawings of the human eye at the 5-month embryo state (a), at birth (b) and as an adult (c). The eye sizes have been normalized to the adult eye size. The actual axial lengths of (b) and (c) are given in Fig. 8.24(a) (after Sivak, 1985).

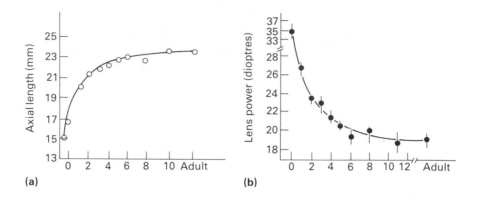

Fig. 8.24. (a) The growth in axial length of the human eye with age. The data were obtained using conventional A-scan ultrasonography techniques (Chapter 7). (b) The change in lens power required to accomodate the change in axial length of the eye (values computed from data given in (a) (after Gordon & Donzis, 1985).

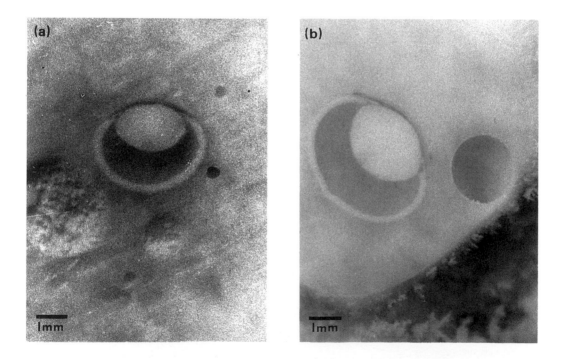

Fig. 8.25. Sagitally hemisected frozen rat eyes at 1 day after birth (a) and as an adult (b). Note increasing sphericity of the lens with age (from Sivak & Dovrat, 1984).

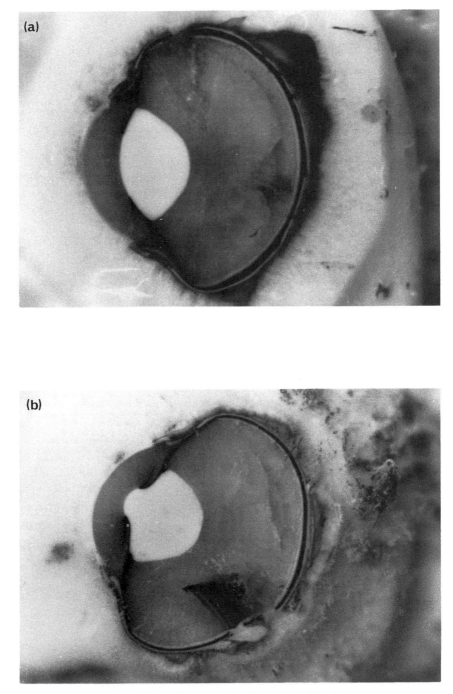

Fig. 8.26. Frozen hemisection of the eyes of a diving duck (Hooded merganser, *Mergus cucullatus*) before (a) and after (b) induced accommodation. The axial length of the eye is 20 mm (from Sivak, 1985).

changing sphere, a shape necessitated by the fact that the cornea has virtually no refractive power in water.

Accomodation under water provides a great challenge to terrestrial animals and certain diving birds meet this challenge by squeezing a portion of the anterior lens hemisphere through the iris (Fig. 8.26). This dramatically increases the power of the lens.

Problem 8.1

In the book *The Lord of the Flies* by William Golding, the boys steal short-sighted Piggy's spectacles and start a fire by focusing the rays of the sun. While this may make good dramatic sense, why is it bad physics?

8.9   Wave theory

It has been possible so far to treat some aspects of optics without assuming any knowledge of the mechanism of the propagation of light. However, such knowledge is essential for an understanding of the principles of microscopy.

The scientific discussion of light began in the 17th century and was led on the one side by Newton who believed that light was made up of tiny corpuscles emitted from the light source like bullets from a gun. For the opposition, Huyghens maintained that light travelled in the form of waves. He also proposed that each point on a wavefront (Fig. 8.27) could act as a source of secondary wavelets and this idea is important for the understanding of the phenomenon of *light diffraction*.

There were two main objections to the wave theory. It was pointed out that light can travel through the vacuum of space whereas sound waves, for example, cannot propagate *in vacuo* and that also while sound waves may travel round corners, it appears, from the strict geometrical shadows of objects, that light cannot.

The first objection has only recently been explained in terms of the dual nature of light (Chapter 9) whereas an explanation for the second was given by Young at the end of the 18th century. He reasoned that the amount of bending of a wave depends on the wavelength and pointed out that waves of long wavelength are more easily bent. For example, if a pipe band disappears round a corner the drum is the last sound to be heard; the higher notes of the bagpipes fade away quicker as the wavelength is smaller. Now as the wavelength of the drum sound is of the order of 0.5 m and the wavelength of light is approximately 0.5×10^{-6} m, the light would be expected to bend much less and so perhaps

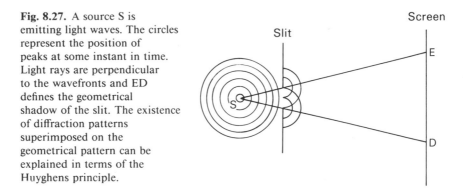

Fig. 8.27. A source S is emitting light waves. The circles represent the position of peaks at some instant in time. Light rays are perpendicular to the wavefronts and ED defines the geometrical shadow of the slit. The existence of diffraction patterns superimposed on the geometrical pattern can be explained in terms of the Huyghens principle.

escape notice. And indeed on close inspection light was found to bend round corners giving rise to diffraction effects at the edge of the geometrical shadow (Fig. 8.28)

Wave motion is defined as the propagation of a disturbance through a medium without the translocation of that medium. Two people, each holding one end of a rope can, by moving the hand holding the rope up and down, transmit vibrational energy from one to the other without the rope moving in a horizontal direction. The waves set up in the rope vibrate perpendicularly to the resting state of the rope and are called transverse waves. *Electromagnetic waves* (light and radio waves also fall into this class) are an example of transverse waves and this type of motion can be expressed mathematically in terms of sine waves and rotating vectors (Fig. 8.29).

The simplest equation to represent a wave mathematically is

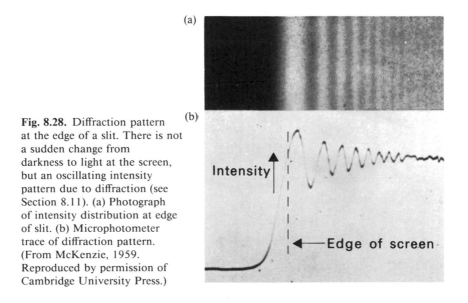

Fig. 8.28. Diffraction pattern at the edge of a slit. There is not a sudden change from darkness to light at the screen, but an oscillating intensity pattern due to diffraction (see Section 8.11). (a) Photograph of intensity distribution at edge of slit. (b) Microphotometer trace of diffraction pattern. (From McKenzie, 1959. Reproduced by permission of Cambridge University Press.)

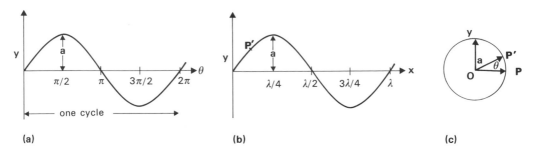

Fig. 8.29. A transverse wave can be represented by either of the equations (a) $y = a$ sin θ or (b) $y = a$ sin $2\pi x/\lambda$. (b) in fact represents a wave of amplitude a and wavelength λ travelling along the x direction. (c) gives the rotating vector representation of wave motion. The vector **OP′** represents the wave at some point in time just as P′ represents the progression of the wave along the x-axis at the same point in time. The magnitude of the vector represents the amplitude of the wave and the angle θ represents the phase of the wave.

$$y = a \sin \theta \tag{8.8}$$

and, for light waves, y could represent the electric field associated with the wave, a is the *amplitude* of the wave, i.e. peak height in Fig. 8.29 (a, b), or the length of the rotating vector in 8.29(c) θ is the *phase* of the wave.

The time taken to complete one cycle is called the *period* and this is the time taken for the wave to travel a distance λ along the x-axis (Fig. 8.29b). It is also the time that the rotating vector takes to complete a revolution, i.e. 2π radians (Fig. 8.29c). Hence we immediately have the relationship between phase angle θ, and distance x

$$\theta/2\pi = x/\lambda \tag{8.9}$$

and so we can write another wave equation

$$y = a \sin \frac{2\pi x}{\lambda}, \tag{8.10}$$

which is equivalent to equation (8.8).

Suppose, however, that we have a wave that does not start at the origin (Fig. 8.30a); then the equation to represent this wave is

$$y_2 = a \sin(\theta - \theta_1) \tag{8.11}$$

and this wave is said to be retarded in phase by θ_1 with respect to the first wave. The rotating vector description of this wave is even simpler (Fig. 8.30b).

If two waves are θ_1 out of phase, i.e. using the sin θ representation of a wave, then they will also be some distance x_1 out of step using the sin $2\pi x/\lambda$ representation. This means that one wave will be moving along the x-axis a constant distance x_1 away from the other wave. The

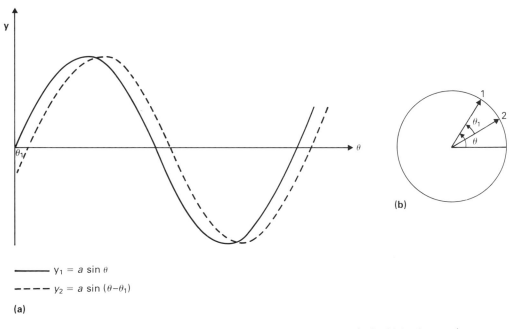

—— $y_1 = a \sin \theta$

- - - - $y_2 = a \sin (\theta - \theta_1)$

(a)

Fig. 8.30. (a) The phase difference between the waves is θ_1. (b) In the rotating vector diagram, the angle between the vectors is θ_1 (a constant) and vector 2 is said to be retarded in phase by θ_1 radians with respect to vector 1.

distance x_1 is referred to as the *path difference* between the two waves and using equation (8.9) it can readily be shown that a phase difference of π radians corresponds to a path difference of $\lambda/2$ while a phase difference of 2π corresponds to a path difference of λ and so on. Of course, when two waves have a path difference of λ they will have come into step with one another and so will be in phase again.

8.10 Interference of light waves

Young first suggested that if light motion was indeed wavelike in nature then it should be possible for two light waves to interact when they arrive together at the same point in space. This reasoning makes good intuitive and mathematical sense. Figure 8.31(a) represent two waves of equal amplitudes and phase and when they interact they add to give a resultant of equal phase, but double the amplitude. Similarly when waves are π radians out of phase they interact to give a zero resultant (Fig. 8.31b). To find the resultant of two waves θ_1 out of phase is cumbersome using the wave equation approach, but is easy when vectors are used (Fig. 8.31c), and so the vector approach will be taken in the remainder of the chapter.

Young devised an experiment to show directly the interference of light waves (Fig. 8.32). Monochromatic light, i.e. light of only one

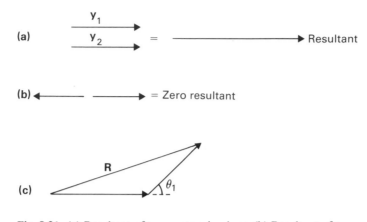

(a)

(b)

(c)

Fig. 8.31. (a) Resultant of two vectors in phase. (b) Resultant of two vectors π radians out of phase. (c) Resultant light intensity is given by $\mathbf{R}^2$. Note the important point that interference patterns cannot normally be obtained from two separate sources as they will be non-coherent. This non-coherence results because the light waves emitted from most sources change phase several million times each second. This change is quite random and so two separate sources cannot be matched for interference purposes. Two coherent sources can, however, be obtained by splitting the light from one source into two separate beams, either by placing two slits in front of the lamp as in Young's experiment, or by using a half-silvered mirror (see Section 8.17).

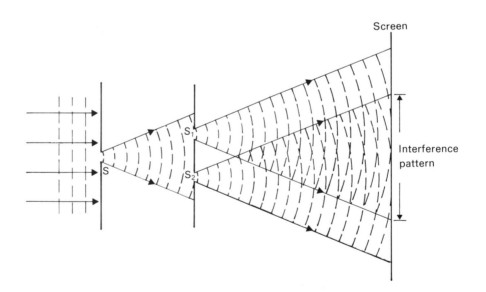

Fig. 8.32. Plane waves arriving at the first slit S give rise to secondary wavelets (diverging rays). The two slits S_1 and S_2 provide coherent sources and so an interference pattern is observed on the screen, set up by the interaction of the secondary wavelets from S_1 and S_2.

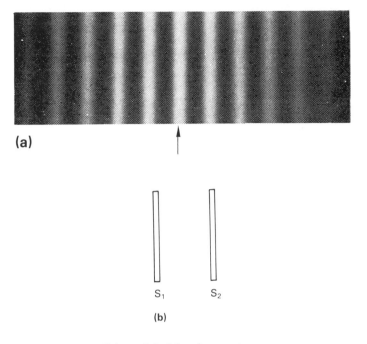

Fig. 8.33. (a) The light and dark bands seen at the screen (Fig. 8.32) are due to the interference of light waves from the slits S_1 and S_2 (b).

wavelength, from a narrow slit first spreads and then passes through two further slits which are close together and parallel to S. Light from these two coherent sources interfere at the screen where a series of narrow light and dark bands can be seen parallel to the slits (Fig. 8.33).

A simplified ray diagram of the experiment may help to show why the alternating bright and dark bands are seen on the screen (Fig. 8.34). Consider a point P on the screen. The waves from S_1 and S_2 have to travel over different distances to reach P and so although they started from S_1 and S_2 in phase they may arrive out of phase at the screen. When the path difference is zero or some integral multiple of λ (phase difference is 2π) then a bright band will be seen, and when the path difference is some odd integral multiple of $\lambda/2$ (phase difference is π) then a dark band is seen. An intermediate intensity (given by R^2 in Fig. 8.31c) will be given for all other path differences.

A phase change can also be introduced between two waves (originally travelling in a medium of refractive index μ_1) by placing a piece of material of thickness t and of different refractive index μ_2 (where $\mu_2 > \mu_1$) in the path of one of the waves (Fig. 8.35). The velocity of the first wave (v/μ_1) is faster than the velocity of the second (v/μ_2) and so the vector representing the second is retarded with respect to the first. How much a wave is retarded obviously depends on μ and on t. The product

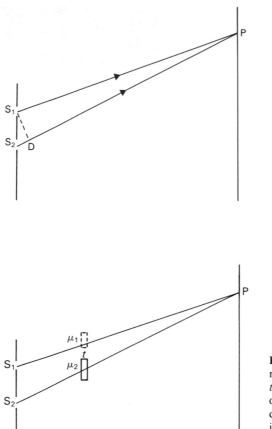

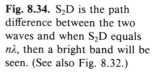

Fig. 8.34. S_2D is the path difference between the two waves and when S_2D equals $n\lambda$, then a bright band will be seen. (See also Fig. 8.32.)

Fig. 8.35. When a material of refractive index μ_2 and thickness t is placed in the path of one of the rays then a further path difference of $t(\mu_2 - \mu_1)$ is introduced.

μt is called the *optical path length*. The difference in optical path length between the two waves is $t(\mu_2 - \mu_1)$. The phase difference introduced is $(2\pi/\lambda)\, t(\mu_2 - \mu_1)$.

Note that if non-monochromatic (white) light is used in Young's experiment then coloured bands are seen.

8.11 Diffraction

Interference fringes are a special case of the more general phenomenon called diffraction which occurs whenever wavefronts are restricted by an aperture or obstacle. Instead of there being a simple geometrical shadow, a complex pattern is obtained (Fig. 8.36) which is best explained by the application of Huyghens' principle.

In microscopy we are concerned primarily with Fraunhofer diffraction which takes place when parallel light impinges on the aperture or obstacle.

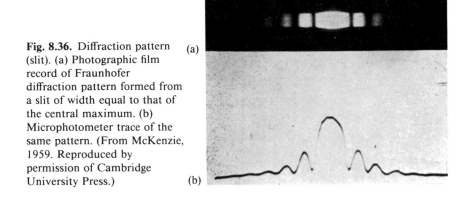

Fig. 8.36. Diffraction pattern (a) (slit). (a) Photographic film record of Fraunhofer diffraction pattern formed from a slit of width equal to that of the central maximum. (b) Microphotometer trace of the same pattern. (From McKenzie, 1959. Reproduced by permission of Cambridge University Press.) (b)

8.12 Fraunhofer diffraction

Lens 1 (Fig. 8.37) produces parallel light which illuminates the slit and lens 2 collects light from the secondary wavelets in the plane of the slit and focuses them on the screen at its focal plane.

Undiffracted light is focused at O and light diffracted through an angle θ is collected at P. SN is drawn perpendicular to the diffracted beam. The waves arriving at P from S and T have a path difference TN. Suppose TN = λ, the wavelength of the light used, then the wavelets from S and from the middle of the slit will have a path difference of $\lambda/2$ and hence will destructively interfere. For every point in the upper half of the slit there will be a corresponding point in the lower half for which the path difference is $\lambda/2$. Thus the first minimum or dark band will occur when TN = λ, and there will be a corresponding minimum on the other side of O. Minima will therefore occur when TN = $\lambda, 2\lambda, 3\lambda$, etc.

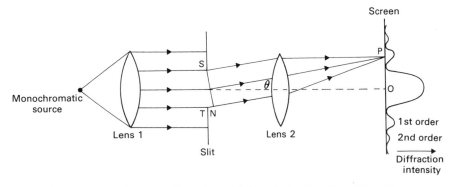

Fig. 8.37. Lens 1 produces parallel light which illuminates the slit and lens 2 collects light diffracted from the slit at an angle θ and focuses it at a point P on a screen placed at its focal plane. The rays from the slit represent light carried by the so-called secondary wavelets (see also Fig. 8.36).

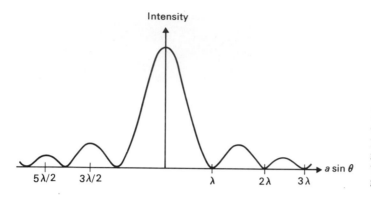

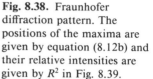

Fig. 8.38. Fraunhofer diffraction pattern. The positions of the maxima are given by equation (8.12b) and their relative intensities are given by R^2 in Fig. 8.39.

There will be a maximum at O as all rays travel equal distances to reach the screen. Consider the case when TN = $3\lambda/2$, then the slit can be divided into three equal parts, the waves from two adjacent bands will annul, but waves from the third part will arrive to give a bright band. This bright band is called the first order maximum.

If ST, the slit width is a, then TN = $a \sin \theta$.

The conditions for minima are (see Fig. 8.38)

$$a \sin \theta = n\lambda \text{ where } n = 1,2,3.... \tag{8.12a}$$

The conditions for maxima are

$$a \sin \theta = \frac{(2n + 1)\lambda}{2} \text{ where } n = 1,2,3.... \tag{8.12b}$$

The position of first minimum is given by

$$\sin \theta = \frac{\lambda}{a}$$

and when θ is small

$$\theta = \frac{\lambda}{a}. \tag{8.13}$$

Equation (8.13) shows that the smaller the slit width, the greater will be the angular separation of the minima.

This is for a rectangular slit and for a circular hole of diameter d it can be shown that

$$\theta = \frac{1.22\lambda}{d} \tag{8.14}$$

Figure 8.39 shows an alternative way of interpreting the diffraction pattern directly in terms of vector theory.

Fig. 8.39. The light arriving at P (Fig. 8.37) from the slit can be considered as n vectors set up by n secondary wavelets and the resultant can be found by adding them head to tail. (a) The phase difference between the vector representing light from the top of the slit 1 and bottom n when they arrive at the screen is 2π and so the resultant is zero. (b) When the phase differences is 3π, a maximum (first diffraction order) is found.

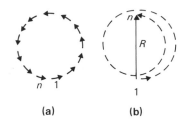

(a)   (b)

8.13 Resolving power of the eye

When a point is being viewed by a magnifying system, e.g. the telescope or the eye, only part of the wavefront from the source passes into the first or objective lens and so diffraction occurs. Instead of seeing a point image one therefore obtains a circular spot of finite extent surrounded by concentric dark and light diffraction rings. If a second point source is viewed, its image will be similar and if it is moved towards the first source a stage will be reached when the diffraction patterns will begin to overlap and the two sources will no longer be resolved. It is almost impossible to give a practical test for the ability to distinguish or resolve a diffraction pattern as being that due to two combined ones and so an arbitrary criterion, proposed by Rayleigh, has been adopted. He said that two objects would be resolved if the maximum of the diffraction pattern from one of the sources coincided with the first minimum from the other.

Light from a distant point object (Fig. 8.40) on the axis of the lens produces an image at I, and light from a distant point object off the axis forms an image I′. The angular separation of I and I′, which are the centres of the diffraction patterns, is equal to the angular separation of

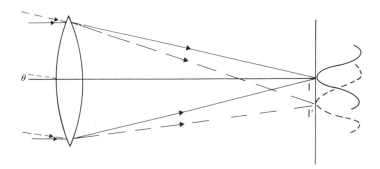

Fig. 8.40. Light from a distant object on the axis of the lines produces an image at I and light from a distant point object off the axis forms an image at I′. According to the Rayleigh criterion the two images are said to be resolved when the maximum of the diffraction pattern from one of the sources coincides with the first minimum of the other.

the distant point objects. Suppose that a circular aperture of diameter d is placed in front of the lens, e.g. the iris of the eye, then the first minimum of the diffraction pattern will occur when

$$\theta = \frac{1.22\lambda}{d} \qquad (8.15)$$

and from Rayleigh's criterion, when the objects are resolved, the maximum of the second source appears at the same place.

The diffraction-limited resolving power of the eye can be computed from equation (8.15). The diameter of the pupil is approximately 5 mm and the wavelength of light to which the eye is maximally sensitive is 500 nm. The eye should therefore be able to resolve two objects when they subtend an angle $\theta = 10^{-4}$ rad at the surface of the eye. However, the experimentally derived resolving power is never greater than 2×10^{-4} rad and so some other factor must be limiting. From the average dimensions of the eye it is possible to calculate the image spacing on the retina that the angular limit of 2×10^{-4} rad implies. As the cornea to retina distance is 2.5×10^{-2} m, then the image separation is 5×10^{-6} m. In the centre of the fovea, where the density of the cone photoreceptors is highest, the cones are about 2×10^{-6} m apart (Fig. 8.13). Hence it appears that the images have to be separated by at least one unexcited cone before they are perceived as separate objects. It is this fact and not diffraction that limits the visual acuity of the eye.

8.14   Diffraction grating

If a diffraction grating containing many slits ruled close together is inserted in place of the single slit in the Fraunhofer diffraction experiment (Fig. 8.41) then a diffraction pattern is observed where the separation between intensity maxima is very large.

Suppose the light rays from all the slits are brought to a focus at P, (Fig. 8.41) then an intensity *maximum* will be observed when the angle θ is such that the phase difference between successive rays is an integral multiple of 2π, i.e. when

$$e \sin \theta = n\lambda \text{ (maximum, cf. single slit)} \qquad (8.16)$$

where e is the distance between slits and n, the diffraction order, takes the values 0, 1, 2, 3, etc. As gratings normally contain over 500 000 lines per metre, i.e. $e = 1/500\,000$ m, then the separation between the diffraction orders is very large. As $\sin \theta$ can be measured readily, the grating is used in physics to determine the wavelength of a source of light.

If white light is used to illuminate a grating then a *spectrum* will be produced at all maxima except the zeroth-order one (in equation 8.16,

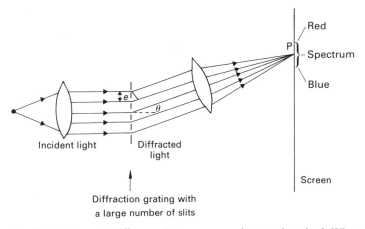

Fig. 8.41. The path difference between successive rays is $e \sin \theta$. When this path difference is λ, the rays from successive slits will constructively interfere and a maximum will be observed. When a white light source is used, a spectrum is obtained at each successive diffraction maximum (except the zeroth).

$\sin \theta$ depends on λ which in the case of white light has a wide range of values). (See also Section 9.3)

Any surface that has lines ruled across it at regular intervals can be used as a diffraction grating and coloured patterns can be seen in the light reflected from insect wing covers for example. These are known as *structural colours*.

Problem 8.2 (after Jarman, 1970)

Electron micrographs of the wing cover of the beetle *Serica sericea* show that it has parallel lines across it and these are 0.8 μm apart. Parallel white light falls perpendicularly on to the surface of the cover and it is viewed at an angle of 45° to the surface. What will be the colour of the wing cover?

(blue: 450 nm, green: 550 nm; red: 650 nm).

8.15 Interference films and filters

Important interference effects arise due to multiple reflections in thin films. For example, soap bubbles or an oil film on water appear brilliantly coloured and the colour depends on the viewing angle. In Fig. 8.42 a thin rectangular film (e.g. soap of refractive index μ) is in a medium of lower refractive index (such as air with $\mu = 1$). If t is the thickness of the film and r the angle of refraction, then the optical path difference (pd) between successive rays focused on the retina is given by

$$pd = 2 \mu t \cos r.$$

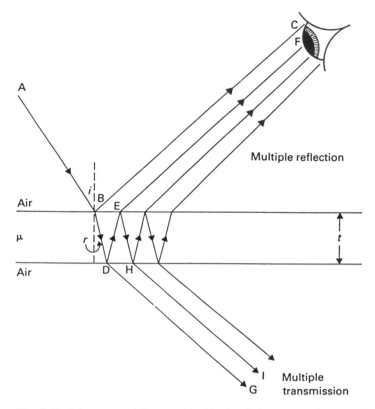

Fig. 8.42. When ray A falls on a thin film it will be partly reflected (BC) and also refracted (BD). The ray BD will in turn be partly reflected at the second interface (DE) and partly refracted (DG). The ray DE will again be partly refracted (EF) and partly reflected. The interference effects due to multiple reflection (BC + EF etc.) and transmission (DG + HI etc.) can be interpreted by considering the optical path differences between successive rays (i.e. ABDEF − ABC etc.) The eye focuses the parallel rays on to one point on the retina and the interference pattern is built up by light rays incident at the angle i.

In the case of normal incidence, the path differences is simply $2\mu t$ and in the limiting case of very thin films, where $t \ll \lambda$, then the path difference is near zero. A very thin film viewed from above in fact appears black, indicating that successive reflected rays undergo destructive interference. This implies that either ray BC or EF has undergone a further phase change of π along its track. It is in fact possible to demonstrate experimentally that rays travelling in one medium suffer a phase change of π when they are reflected at the interface with a denser medium. The phase change of π (equivalent to a path difference of $\lambda/2$) therefore occurs at the first reflection (ABC) and not at the second (BDE) and so in the limiting case, for very thin films, successive rays reflected from the film undergo destructive interference and the film appears black. When μt increases to near the wavelength of visible light, then destructive interference occurs for certain wavelengths (e.g. $2 \mu t = 600$ nm $=$ red light) and the film, when viewed with white light, will appear as white minus red, i.e. green in colour. Light passing through the film does not suffer the phase change and so in very thin films the transmitted light will appear white, while in thicker films ($2 \mu t = 600$ nm for example), the transmitted light will have a red hue. The relationship between thickness and colour in reflected light is in fact used to estimate t for thin sections prepared for the electron microscope.

By forming successive thin films on top of one another, with slightly different refractive indices, then multiple interference patterns are produced. Multiple layer interference patterns are responsible for the structural colours in fish scales and around the eye of the squid. When thin films are deposited on the surface of a mirror in a highly controlled fashion, it is possible to produce interference filters that reflect a high percentage of certain spectral ranges while others are almost completely transmitted. These *dichroic mirrors* form an integral part of modern fluorescence microscopes (see Section 9.12).

8.16 Microscopy

Great advances have been made in biology with various types of microscope ranging from Hooke's original simple microscope to the modern electron microscope. In a compound microscope (Fig. 8.43) the objective forms an image of the object and this light is then imaged by the eye piece and the eye to form a virtual image beyond the objective. As in the case of the telescope, diffraction takes place in this system, and

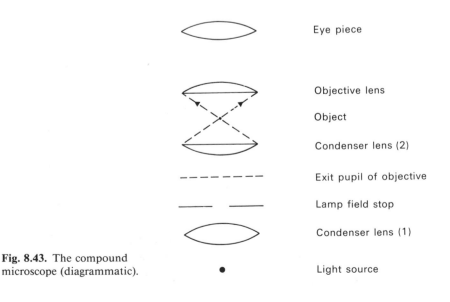

Fig. 8.43. The compound microscope (diagrammatic).

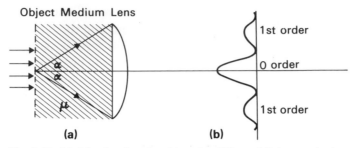

Fig. 8.44. (a) After leaving the object the diffracted light travels through a medium of refractive index μ to the objective lens of angular aperture α. (b) Diffraction pattern at objective lens. According to Abbé's criterion the aperture of the objective must be wide enough to collect at least the zeroth and both first order diffraction patterns.

limits the resolution, only here light is scattered and diffracted by the object. For the purpose of estimating the resolving power of the system, the object can be considered as a diffraction grating. Abbé proposed the now accepted criterion for microscope resolution. He assumed that for faithful reproduction of an object by a microscope the aperture of the objective must be wide enough to transmit at least the zeroth and both first order diffraction patterns. In fact, the more of the diffraction pattern that is lost, the poorer will be the production of the image. Suppose the object is a grating of element spacing e (Fig. 8.44) and suppose at first that the condenser lens (2) is omitted so that parallel light falls on the object, then in this case the position of the first order maximum is given by

$$e \sin \alpha = \lambda' \tag{8.17}$$

where λ' is the wavelength in the medium between the object and the lens, i.e.

$$e \sin \alpha = \frac{\lambda}{\mu}$$

where λ is the wavelength of the light in air, or

$$e = \frac{\lambda}{\mu \sin \alpha}. \tag{8.18}$$

For e, in this case the limit of resolution, to be small, α has to be as large as possible, i.e. the objective must be placed as near the object as possible. It is not always possible to illuminate with parallel light as too much light is lost from the source in illuminating a large area of the slit. A condenser lens (lens 2 in Fig. 8.43) is therefore used to focus light on to the object and this arrangement not only increases the image brightness but also the resolving power of the microscope as now the resolution equation becomes

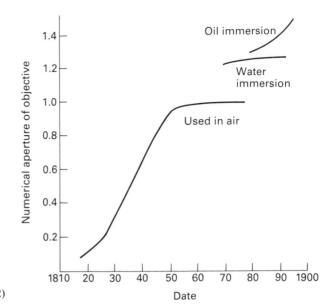

Fig. 8.45. Increase in numerical aperture of microscope objectives in the 19th century. (After Turner, 1972)

$$e = \frac{\lambda}{2\mu \sin \alpha} \qquad (8.19)$$

a fact that can only be explained by complex mathematical reasoning. Apart from making α as large as possible, the resolution can also be improved by making λ as short as possible and μ as big as possible by using oil-immersion lenses. The ultraviolet microscopic has improved resolution but suffers in that special glass has to be used in the lens construction and also the images must be recorded photographically.

In equation (8.19) $\mu \sin \alpha$ is called the *numerical aperture* (NA) and is marked on microscope objectives together with the magnification.

The early 19th century microscope manufacturers realized empirically that the wider the cone of light from the specimen entering the front lens of the objective, the better the fine detail which could be seen. The largest theoretical angle for α is 90°, but in practice there must be some space between the specimen and the lens (the 'working distance'), so the maximum possible angle is about 85° and this was achieved by 1860. The improvements thereafter come from the use of water and oil immersion lenses (Fig. 8.45).

8.17 Interference microscopy

The interference and phase-contrast microscopes are of great value to microscopists as they enable small objects to be viewed without the prior use of staining techniques. As there is usually only a small

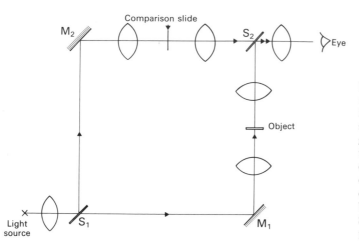

Fig. 8.46. The interference microscope. The light is split into two beams by the half-silvered mirror S_1, one-half going through the object and the other half going through the comparison slide which consists of a wedge of variable optical path length. The two beams are then recombined by S_2 and so interference is possible. M_1 and M_2 are mirrors.

difference in absorption between, say, a cell and its bathing medium, there will be little difference in amplitude between the light which passes through the medium and that which passes through the specimen. There will, however, be a phase difference introduced because of a refractive index difference between the specimen and medium. Because the eye cannot detect differences in phase, the purpose of the interference and phase microscopes is to convert this phase difference into an amplitude difference. One way of doing this is to use round-the-square interference (Fig. 8.46).

The observed image arises from the sum of two light beams. One passes through the object and the other traverses a comparison slide by means of which the phase and amplitude of the comparison beam can be adjusted. In Fig. 8.47 the vector **OB** represents the background light and **OT** the light transmitted through the object. As the latter has a slightly higher refractive index than the suspending medium, **OT** will lag behind **OB** by a small angle θ. The phase of the light passing through the comparison slide is now adjusted until the object appears dark. The phase of the comparison beam will have been advanced relative to **OB** by an angle $180° - \theta$ (Fig. 8.48). The object appears dark as the two vectors **OC** and **OT** are equal in amplitude but $180°$ out of phase. The background does not appear dark, however, as the interaction of the two vectors representing the background and comparison light produces a resultant **R**. Contrast is therefore achieved and the object appears dark in a bright surround.

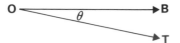

Fig. 8.47. See text for explanation.

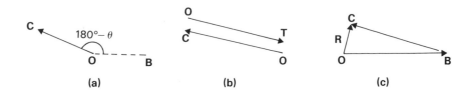

Fig. 8.48. (a) **OC** represents the amplitude and phase of the light which has passed through the comparison slide and its phase has been adjusted so that it is exactly out of phase with light transmitted through the object **OT** (b). The background does not appear dark, however (c) as the background vector **OB** and comparison vector are not 180° out of phase.

8.18 Phase microscopy

The phase contrast microscope was invented by Zernicke and contrast in this system is achieved by interference between undiffracted light and light diffracted from the object. The phase plate (Figs 8.49 and 8.50) and annular diaphragm lie in the focal plane of the objective and the condenser lens respectively. In the absence of an object, therefore, light which passes through the annulus falls on the etched ring on the phase plate.

When an object is present, light is diffracted. The vector **OB** (Fig. 8.51) represents the background light, and **OT** the light transmitted through the object. **OT** can in fact be split into two components, **OB'**, the undiffracted light (zeroth order) which has the same phase and almost the same amplitude as the background light and **B'T** which represents the diffracted beam. When θ is small, i.e. the refractive indices of object and medium are similar, then **B'T** is very nearly $\pi/2$ rad out of phase with **OB'**. Undiffracted light passes through the etched

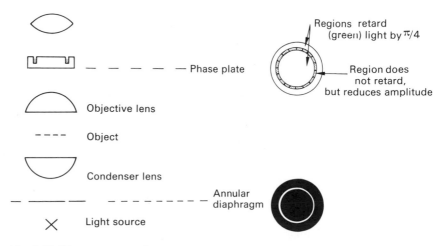

Fig. 8.49. Phase contrast microscope.

156 CHAPTER 8

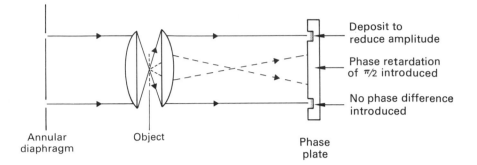

Fig. 8.50. A purely diagrammatic representation of the path of the light rays in the phase contrast microscope ———— background and zeroth order diffraction; – – – – – first, second, etc, order diffraction.

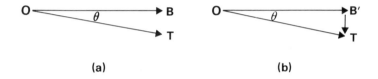

(a)   **(b)**

Fig. 8.51. (a) **OB** represents the background light and **OT** the light transmitted through the object. (b) **OT** can be resolved into two components, **OB′** and **B′T**; the former has the same phase and almost the same amplitude as the background light and corresponds to the zeroth order of diffraction. **B′T** represents the first, second and other orders and is $\pi/2$ out of phase with **OB**.

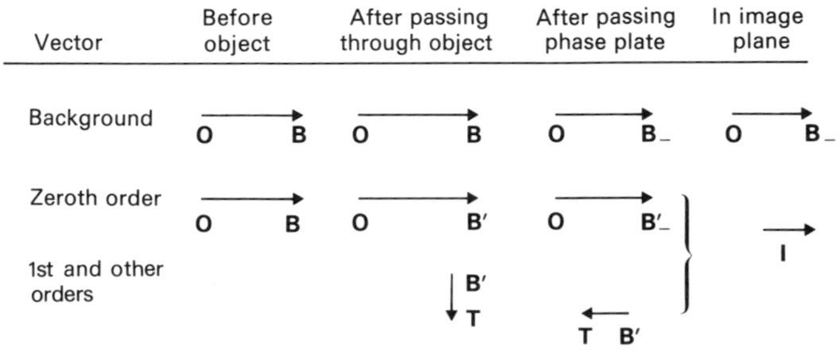

Vector	Before object	After passing through object	After passing phase plate	In image plane
Background	O———→B	O———→B	O———→B_	O———→B_
Zeroth order	O———→B	O———→B′	O———→B′_	→I
1st and other orders		↓B′ ↓T	←T B′	

Fig. 8.52. Vector representation. B_ indicates that the background light has been reduced in amplitude on passing through the phase plate.

region of the phase plate and is unchanged in phase but reduced in amplitude **OB_** (Fig. 8.52). The diffracted light, however, strikes the plate at other points and is further retarded in phase by $\pi/2$ rad.

The zeroth and other orders combine together in the image plane to form the image. The resultant vector **I** is reduced in length compared to the background light **OB_** and the object thus appears dark on a bright background. This is called *positive phase microscopy*.

Note: If there were no annulus then undeviated light would also suffer a phase change and so the system would not work.

8.19 Polarization optics

Light is an electromagnetic radiation consisting of electric **E** and magnetic **H** vectors vibrating at right angles to each other and to the direction of propagation. By convention, polarization is associated with the electric vector.

In light emission from a group of atoms, the direction of vibration of the **E** vector changes about every 10^{-8} s and unpolarized light travelling at right angles to the plane of the paper can be represented by Fig. 8.53.

There are some special crystals that will pass light vibrating in only one direction. Tourmaline is an example of this *dichroic* class of crystals (Fig. 8.54).

Only light with electric vectors vibrating parallel to the long edges of the crystal passes through. Light is said to be *plane polarized* after passing through the first crystal, the *polarizer*, and this light will be stopped by a similar crystal, the *analyser* placed at right angles to the polarizer. As tourmalines are coloured, they are of little use in optics.

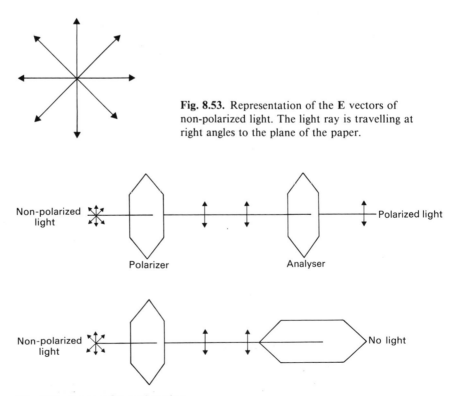

Fig. 8.53. Representation of the **E** vectors of non-polarized light. The light ray is travelling at right angles to the plane of the paper.

Non-polarized light — Polarizer — Analyser — Polarized light

Non-polarized light — No light

Fig. 8.54. See text for explanation.

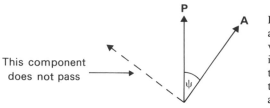

Fig. 8.55. The polarizer and analyser allow only light vibrating along the axes indicated to pass through. Only the component **P** cos ψ passes through both polarizer and analyser.

Similar effects can, however, be produced by polaroid films which are not coloured. These are formed from films of polyvinyl alcohol which are first stretched to line up the polymers and then impregnated with iodine. Polaroid films are strongly dichroic and allow a small but useful range of wavelengths to pass. In the polarizing microscope, the light passes through a polaroid polarizer before passing through the object and after the object the light goes through the analyser before reaching the eye. Normally the angle ψ between polarizer and analyser is under the control of the observer (Fig. 8.55).

It is only possible to get zero light intensity if the polarizer and analyser are at right angles to each other. This is because any vector can be resolved into two components (Fig. 8.55). Light is plane-polarized after passing the polarizer and if it has amplitude **P**, then a component **P** cos ψ will pass through the analyser. Hence if $\psi = 0$, i.e. the orientation of polarizer and analyser coincide, then all the polarized light gets through. If the axes are at right angles then no light emerges.

There are some crystals which, while not being completely dichroic, do slow down the movement of light polarized in one plane differently from that polarized in a different plane. This property is called *birefringence*, and it is shared by many arrangements of molecules and cells in living systems. We can attempt to explain this property as follows.

Consider the case of a diatomic molecule, containing atoms A and B (Fig. 8.56). When the light is vibrating parallel to the line joining the two atoms (Fig. 8.56a) the polarizing effect of the light on one atom is

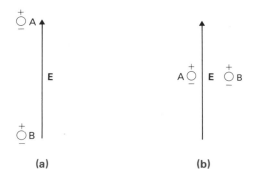

Fig. 8.56. (a) Interaction between the **E** vector and the dipoles is high. Direction AB defines the major axis. (b) The direction perpendicular to AB defines the minor axis.

enhanced by the induction of the other polarized atom. Dipole A enhances the polarization of dipole B and the interaction between the light and the molecule AB is high. The refractive index is a measure of this interaction. When, however, the molecule is turned through 90° (Fig. 8.56b) each dipole opposes the polarization of the other and interaction is low. In a fibre in which the chain of moleclues lie parallel to each other and to the length of the fibre, it would therefore be expected that the refractive index for light vibrating parallel to the fibre length would be very much greater than for light vibrating at right angles. In a plant cell wall, for example, the direction of the cellulose chains (i.e. the direction of the *major axis* of refractive index) can be determined even in the smallest microscopically visible piece of wall with the aid of the polarizing microscope.

8.20 The polarizing microscope

Many biological tissues and even individual cells have some of their constituent molecules arranged in an ordered fashion and so they are suitable subjects for examination by the polarizing microscope. Any conventional microscope can be converted to polarizing optics simply by inserting one piece of polaroid film (polarizer) between the light source and the stage and another piece (analyscr) between the stage and the eye-piece. Specialist polarizing microscopes have a built-in rotating stage. The following experiment will demonstrate the method of operation of the microscope.

Set up a polarizing microscopic with the polarizer and analyser at right angles, place a small piece of cell wall, e.g. from the algal cell *Nitella*, on the stage and illuminate with white light. When the stage is rotated, four positions of darkness are found (when the long axis of the

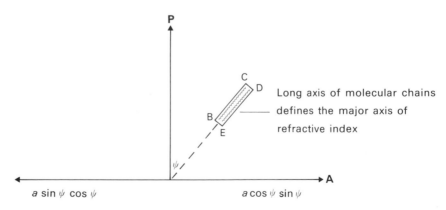

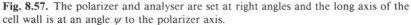

Fig. 8.57. The polarizer and analyser are set at right angles and the long axis of the cell wall is at an angle ψ to the polarizer axis.

plant material is parallel to **P** and then parallel to **A**) showing that the cell wall is birefringent.

Suppose the long axis of the wall is at an angle ψ to the polarizer; if light on passing the polarizer has amplitude **a**, the component along the direction BC will be **a** cos ψ and this is the amplitude of the light which will pass through the wall of thickness d. On emerging from the wall, a component of this light will pass through the analyser. The final amplitude of this light will be **a** cos ψ sin ψ and the vector will be towards the right.

Remembering that light which has been plane-polarized by the polarizer also has a component along the direction EB, a similar calculation shows that the amplitude which finally passes through the analyser is **a** sin ψ cos ψ and the vector is towards the left.

From the figure as it stands, the two vectors have equal amplitudes, but are opposite in direction and so will give intensity zero on reaching the eye. However, this is only true if the cell material has equal refractive indices for light vibrating along EB and BC. In the case of the plant cell wall, the two indices are different (μ_1 in one direction and μ_2 in the other) and so phase difference is introduced between the two vectors. The path difference is $(\mu_1 - \mu_2)d$, where d is the thickness of the cell wall, and the phase difference is $(2\pi/\lambda)(\mu_1 - \mu_2)d$. When the path difference introduced is an integral number of some wavelength λ, then light of that wavelength will be missing and the specimen will appear coloured against a black background.

Suppose that the path difference introduced by the specimen is 536 nm, the wavelength of green light, then if white light is used, green will be missing and so the light will appear red. If the path difference is less than 536 nm, the wall will appear yellowish and if it is greater, it will appear green or blue. The scale which associates path difference with colour is called *Newton's colour scale* (Table 8.1) and can be used to decide which are the directions of the major and minor axes of the wall and hence the orientation of the cellulose molecules comprising the wall.

In order to obtain the direction of the major axis of *Nitella*, for example, a piece of birefringent material, with its axes clearly marked, is

Path difference (nm)	Colour of birefringent material between crossed polaroids	
0	Black	
330	Yellow	↑ Subtraction colours
536	Red	— First order red
575	Violet	↓ Addition colours
747	Green	

Table 8.1. Newton's colour scale.

now inserted into the polarizing microscope between the polarizer and analyser and at an angle of 45° to the polarizer. The slice of crystal normally used introduces a path differences of 536 nm between light traversing its major and minor axes and so in this case the field of view in the microscope appears red. The colour of the plant cell wall depends on its orientation relative to the axes of the red plate.

When the long axis of the wall is parallel to the major axis of the plate, the wall appears green, which is an *addition colour* in Newton's series. This means that the path difference introduced by the plate adds to that introduced by the wall, and hence the two major axes coincide when the long axis of the wall is parallel to the major axis of the plate. This means that the cellulose molecules run parallel to the long axis.

This conclusion can be checked by rotating the wall through 90°. It now appears yellow, which is a *subtraction colour* and from this we can conclude that, as expected in this orientation, it is the minor axis of the wall that is parallel to the major axis of the plant. The cell wall is said to be *positively birefringent* as the morphological long axis coincides with the major birefringent axis.

Similarly it can be shown that muscle fibres are positively birefringent, whereas myelinated nerve cells are negatively birefringent. If the slide holding the nerve is flooded with chloroform (a lipid solvent) then

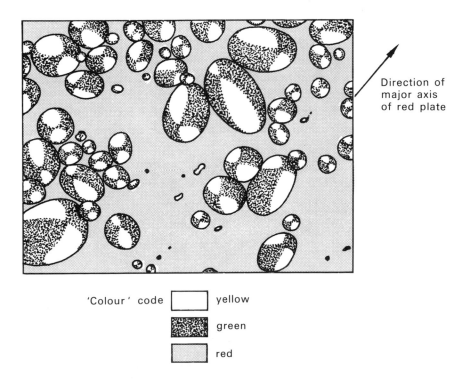

'Colour' code ☐ yellow

▓ green

☐ red

Fig. 8.58. Maltese cross appearance of starch grains (after Preston, 1952).

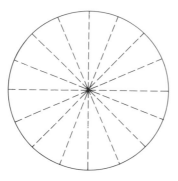

Fig. 8.59. Conclusion from the experiment shown in Fig. 8.58. The molecules comprising the starch grain are radially oriented.

the sign of the birefringence changes showing that it is set up in the natural state by the radially oriented phospholipids (see Section 9.3).

One of the most beautiful examples of the use of the polarizing microscope is in the observation of a scraping from a cut potato. The starch grains have a Maltese cross appearance (Fig. 8.58) and the arms of the cross parallel to the major axis of the plate show an additional colour, green, while the arms at 90° to this direction are yellow. From this we can conclude that the starch molecules are radially oriented (Fig. 8.59).

8.21   Birefringence and polarization by double refraction

Many materials with a crystalline structure are doubly *refracting* or *birefringent*. This implies that the velocity of light of a wave propagating through the crystal is not the same in all directions.

Experimentally it is found that if a ray of light falls on a birefringent crystal, then the ray is split into two rays. If the crystal is rotated through 360° with the incident ray as the axis of rotation, then one of the rays is undeviated, while the second ray rotates around it (Fig. 8.60). The first ray is termed the *ordinary ray* and the second the *extraordinary ray*. Snell's Law holds for the ordinary ray but not the extraordinary ray as the velocity of the latter is different in different directions.

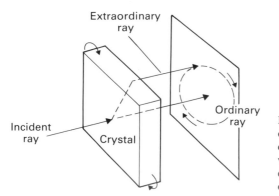

Fig. 8.60. A ray of light falling on a birefringent crystal can be split into two (ordinary and extraordinary) rays. When the crystal is rotated with the incident ray as the axis of rotation the extraordinary ray rotates round the stationary ordinary ray.

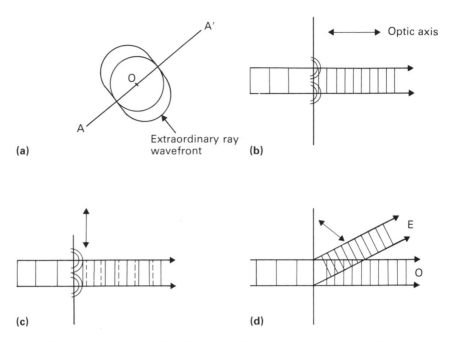

Fig. 8.61. (a) Spherical and ellipsoidal waves diverge from a point source in a uniaxial crystal. (b) In the direction of the optic axis there is no distinction between the ordinary (O) and extraordinary (E) rays. (c) Perpendicular to the optic axis there is a maximum difference between the velocity of the O- and E-ray, i.e. they travel in the same direction but are out of step. (d) At an arbitrary angle to the optic axis the O- and E-rays differ in direction and velocity (after Sears & Zemansky, 1964).

The wavefront for a wave propagating from O in such a crystal would consist of a sphere for the ordinary ray and an ellipsoid for the extraordinary (Fig. 8.61). The direction in which the velocities of the ordinary and extraordinary rays are equal is the *optic axis* of the crystal (AOA′) and crystals with only one axis are termed *uniaxial*. The velocities of the two rays are maximally different at right angles to the optic axis.

It is found experimentally that the ordinary and extraordinary light waves are linearly polarized in mutually perpendicular directions and, as the rays can be separated from one another, uniaxial crystals provide a powerful means of obtaining linearly polarized light. The Scottish physicist Nicol was one of the first to develop a device to produce strong linearly polarized light.

The *Nicol prism* (Fig. 8.62) consists of two wedges of Icelandic spar, originally cut from the same crystal and then cemented together along the cut edge with Canada balsam. The two rays are separated in the first prism where the ordinary ray undergoes total internal reflection while the extraordinary ray passes to the second wedge and emerges linearly polarized.

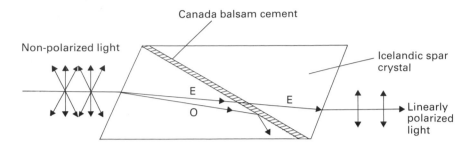

Fig. 8.62. The Nicol prism is produced from two segments of Icelandic spar cemented with Canada balsam. The refractive index of the latter is such that the ordinary ray is totally internally reflected.

Another useful optical device is the Wollaston prism (Fig. 8.63) in which the two wedges are cemented together such that the optic axes of the two pieces are mutually perpendicular. The optic axis of the first wedge is at right angles to the incident light ray so that the ordinary and extraordinary rays travel in the same direction through the wedge, but their velocities are very different. They are incident at an angle to the second wedge and the ordinary ray will be refracted according to Snell's Law whereas the extraordinary ray will not. The paths of the ordinary and extraordinary rays will therefore no longer be the same and, in fact, in this geometrical arrangement they are maximally different. This prism forms the basis for the Differential Interference Contrast microscope.

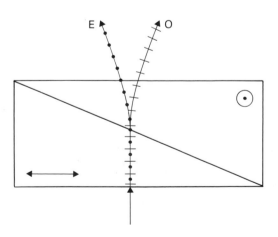

Fig. 8.63. The Wollaston prism is produced from two calcite or quartz prisms cemented together such that their optic axes are mutually perpendicular.

8.22 Differential Interference Contrast (DIC) or Nomarski microscopy

The Nomarski interference method for producing images is basically a combination of polarization and interference microscopy techniques. The microscopic comprises a polarizer and analyser and two special optical devices (Wollaston prisms), each composed of two birefringent prisms cemented together. The optical axes of the birefringent prisms are at right angles to each other. The polarizer is set so that its axis is at 45° to the optic axis of the first birefringent prism. Light from the polarizer can be considered to be split into ordinary and extraordinary

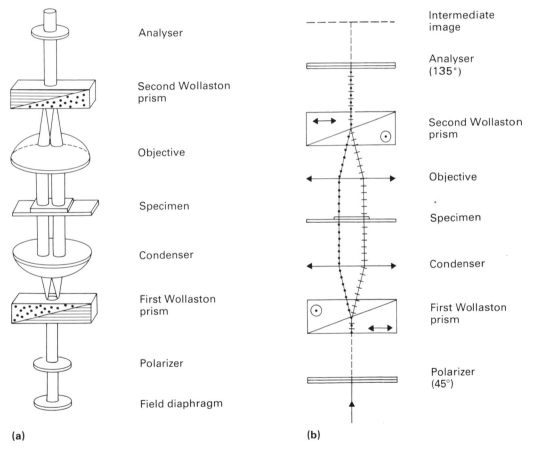

(a)   (b)

Fig. 8.64. (a) Diagrammatic representation of a double-beam (Nomarski) interference contrast microscope. (b) Ray diagram representation of a DIC microscope. Light emerges from the polarizer with a plane of vibration at 45° to the plane of the paper (----). The ordinary (++++) and extraordinary (••••) rays are separated and take different paths through the specimen. The two rays are recombined in the second (inverted) Wollaston prism. The analyser is set at 90° to the polarizer so that any phase difference introduced between the ordinary and extraordinary waves by the specimen is converted to an amplitude difference in the final image (as in the polarizing microscope, Section 8.20) (after Lang, 1970).

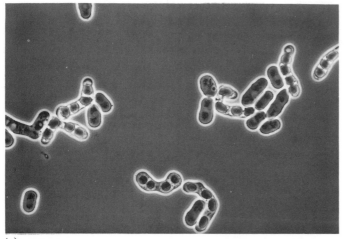

(a)

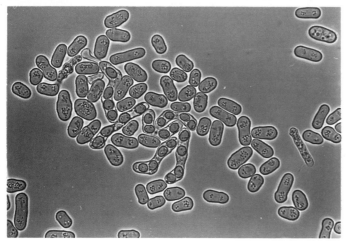

(b)

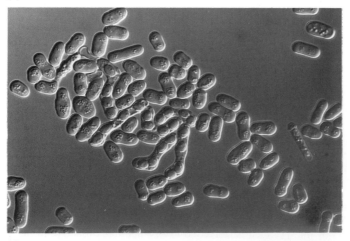

(c)

Fig. 8.65. Yeast cells viewed with three different types of microscope optics: (a) conventional compound optics (Section 8.16); (b) phase optics (Section 8.18); and (c) differential interference contrast optics (Section 8.22). (Photographs kindly provided by Linda Barnett and used with permission.)

rays in the first prism (Fig. 8.64). They will take the same path, but one will be retarded with respect to the other as one vector is vibrating along the optic axis and the other at right angles to it. On passing into the second prism, the ordinary ray is bent towards the base and the extraordinary ray is bent away from it. In this way a separation of the beams is achieved. On passing through the specimen the optical path lengths of the two beams can be different and hence when the two beams are recombined after passing through a second, matched Wollaston prism, then the interference pattern produced varies according to this difference in optical paths (Fig. 8.64). If white light is used the field of view will appear coloured. When monochromatic light is used single interference fringes can be seen in the field of view without a specimen in place. The light intensity varies across the width of a fringe and it is this gradient which results in a three-dimensional *appearance* in a strictly two-dimensional image (Fig. 8.65).

8.23   Comparison of different types of light microscopy

The three photographs in Fig. 8.65 show yeast cells at the same magnification. In conventional and phase microscopy the halo (from light scattered at the abrupt refractive index change) is very apparent at the cell–medium interface. The internal structures of the cells including the ascospores are resolved under both conventional and phase optics while cell texture is more apparent with Nomarski optics (Fig. 8.65c).

The three photographs shown are used, along with other information, to help in identifying yeasts (Barnett *et al.*, 1983).

References

Barnett, J.A., Payne, R.W. & Yanon, D. (1983) *Yeasts: Characteristics and Identification.* Cambridge University Press.

Duncan, G. (1989) The living lens. *Biological Sciences Review*, **1**, 7–11.

Fernald, R.D. & Wright, S.E. (1983). Maintenance of optical quality during crystalline lens growth. *Nature*, **301**, 618–20.

Gordon, R.A. & Donzis, P.B. (1985) Refractive development of the human eye. *Archives Ophthalmology*, **103**, 785–9.

Jarman, M. (1970) *Examples in Quantitative Zoology*. Arnold, London.

Lang, W. (1970) Nomarski differential interference — contrast microscopy. *Reprint 541–210–2–5–e.* Zeiss Information, Oberkochen, W. Germany.

McKenzie, A.E.E. (1959) *A Second Course of Light*. Cambridge University Press.

Preston, R.D. (1952) *The Molecular Architecture of Plant Cell Walls*. Chapman & Hall, London.

Sears, F.W. & Zemansky, M.W. (1964) *University Physics*. Addison-Wesley, Reading, Mass.

Sivak, J.G. (1985) Optics of the crystalline lens. *American Journal of Optics and Physiological Optics*, **62**, 299–308.

Sivak, J.G. & Dovrat, A. (1984) Early postnatal development of the rat lens. *Experimental Biology*, **43**, 57–65.

Turner, G.L'E. (1972) Micrographia Historica: the study of the history of the microscope. *Proceedings of the Royal Microscopical Society*, **7**, 120–49.

Wald, G. (1950) Eye and camera. In *From Cell to Organism*. Freeman, San Francisco.

Further reading

Bradbury, S. (1976) *The Optical Microscope in Biology*. Arnold, London.

Campbell, I.D. & Dwek, R.A. (1984) *Biological Spectroscopy*. Benjamin Cummings, Menlo Park, USA.

Duncan, G. (1981) *Mechanisms of Cataract Formation in the Human Lens*. Academic Press, London.

Gregory, R.L. (1972) *Eye and Brain*. Wiedenfield & Nicholson, London.

Jenkins, F.A. & White, H.E. (1957) *Fundamentals of Optics*. McGraw-Hill, New York.

Setlow, R.B. & Pollard, E.C. (1962) *Molecular Biophysics*. Pergamon Press, London.

Wearden, T. (1981) Technology and application of fibre optics. *Electronic Product Design*, October, 37–41.

9 Quantum Optics: Interaction of Energy with Matter

9.1 The photoelectric effect

We have seen much evidence which points to light energy being carried from point to point by wave motion. We have also been able to describe certain of the interactions of light with matter, e.g. diffraction, solely in terms of the wave theory. However, a crucial phenomenon called the *photoelectric effect* was found to be a stumbling block to a complete acceptance of the wave theory.

It was found that when light falls on a photosensitive metal plate (Fig. 9.1), electrons are emitted and can be collected at a non-photosensitive anode. With a steady beam of light falling on the so-called photocathode, a steady current is observed to flow round the circuit. The current can be explained if it is imagined that the light energy excites the electrons so that they gain sufficient energy to leave the surface of the photocathode.

The velocity or, more correctly, the kinetic energy with which the electrons leave the surface can be measured by reversing the polarity of

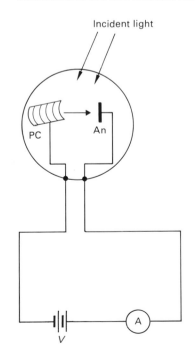

Fig. 9.1. In the dark, no current is registered on the ammeter A, but when light falls on the photocathode PC, electrons are driven off and are attracted to the anode An because of the voltage difference V between cathode and anode.

169

the electrodes (see Chapter 10 for an explanation of the electrical terms used here) so that electrons are attracted back towards the photocathode. It is found experimentally that there is a certain potential V_0 at which no current flows, and V_0 gives a measure of the energy of the electrons, i.e. the higher the energy the greater will have to be V_0 to keep the current at zero.

From wave theory we might expect that the higher the intensity of light (the greater the amplitude of the waves) the greater V_0 should be. However, experimentally V_0 is found to be independent of intensity. V_0 varies only as the frequency of the light used, and the higher the frequency the greater is the energy of the emitted electrons.

This observation led Einstein to postulate that light energy is carried by small discrete packets, now called *quanta* or *photons*. The energy of a quantum E is given by the relationship

$$E = hv \tag{9.1}$$

where v is the frequency of the light and h is *Planck's constant* which has the value 6.6×10^{-34} Js. According to the quantum theory, increasing the intensity only means increasing the *number* of electrons per second leaving the photocathode. This has no effect on the voltage required to reduce the current to zero. The higher the frequency of the light, however, the greater is the energy of the incident photon and the greater will be the kinetic energy of the electron driven off by the photon.

9.2   Light emission

It is now well known, through the work of Lord Rutherford and others, that an atom consists of a positively charged central nucleus with electrons in orbit around it (Fig. 9.2). The basic assumptions of the quantum theory are:

1   The electron orbits are such that the electrons radiate no energy while they are in them and the atom is in a low energy, non-radiating state, often called the *ground state*, when all the orbitals have their prescribed number of electrons.

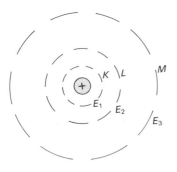

Fig. 9.2. The coulombic interaction (Chapter 10) between the nucleus and innermost orbitals is high and so the orbitals have a high energy content $E_1 > E_2 > E_3$. When electrons are lost from the innermost orbitals, labelled K, L, M, etc., a very large potential energy hole is formed and so high energy photons are emitted when electrons jump down from the outer orbitals.

2 The excited state is achieved by the absorption of sufficient energy to remove one or more of the electrons from a given orbital; a potential energy hole is thus created and the closer the electron to the nucleus, the deeper the hole. The atom loses this excess energy in quantal form when an outer electron jumps down to fill a hole in an inner orbital. The energy of the emitted photon is in fact the difference in energies of the two orbitals involved, i.e.

photon energy $= \Delta E$

and as

$c = \nu\lambda$

where c is the velocity of light then

$\Delta E = hc/\lambda$

or

$\lambda = hc/\Delta E.$ (9.2)

When transitions are between the outermost orbitals, the energy involved is small and the photons involved are in the visible range. When innermost orbitals are involved, ΔE is large and the photons are called X-rays.

9.3 Investigation of biological ultrastructures by X-ray diffraction

Theoretical background

The amount of information that can be derived from an examination of any material depends ultimately on how fine a probe is used. For example, an examination of biological tissues by the optical microscope (Chapter 8) is limited by the wavelength of visible light which is in the region of 500 nm.

Details of molecular arrangements within the tissue are not resolved as they are only 1–10 nm apart and so for these another technique has to be used where the probe is much finer. The wavelength of X-rays is in the region of 0.1 nm and so they might appear to provide an excellent probe. Unfortunately, however, no substance has yet been found that can focus X-rays in the way that a lens gathers together the diffraction patterns from an illuminated object and assembles them to form an image. In the case of X-rays we are left with the diffraction pattern which, at its simplest, we record on film. X-rays analyses are limited to regularly repeating structures as it is only from the constructive interference of X-rays scattered from several identical structures that a

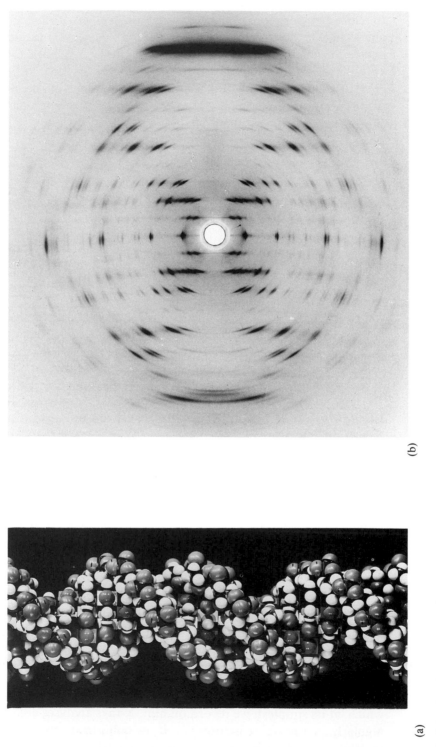

(b)

(a)

Fig. 9.3. (a) Molecular model of DNA together with the diffraction pattern (b) obtained from DNA crystals. Note the helical arrangement of the molecules and the cross-wise pattern of the dark spots on the diffraction picture. (Photographs kindly provided by Professor M. H. F. Wilkins, FRS.)

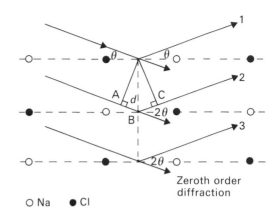

Fig. 9.4. Representation of the scattering of X-rays from repeating units (planes distance d apart) in a sodium chloride crystal. Rays 1 and 2 are reflected from identical planes and when they interfere at the plane of the film a diffraction pattern is produced. If the path difference between 1 and 2 is exactly equal to λ, the wavelength of the X-rays used, a first order diffraction spot is produced. Note that the angle between the incident and reflected rays is 2θ,

strong diffraction pattern is derived.

In order to understand these diffraction patterns some mathematical insight is required. For example, Crick, of the Crick and Watson team, realized that the crosswise pattern on the X-ray diffraction picture from DNA (Fig. 9.3) was due to a repeating helical structure only because he had previously worked out a mathematical treatment of helix diffraction from a totally different system.

The foundation for the mathematical treatment of diffraction patterns was laid down by Sir Lawrence Bragg. He suggested that instead of considering each atom to act as a scattering point for the X-rays, the planes of atoms that make up a repeating structure should be considering as *reflecting units* for the X-rays. The X-ray diffraction pattern would then be made up of all rays reflected from the different planes. The NaCl crystal for example consists of a cubic array of sodium and chloride ions and Fig. 9.4 is a two-dimensional representation representation of this crystal. The X-rays can be considered as being reflected from identical planes. The path difference introduced between ray 1 and ray 2 will be AB + BC and if θ is the glancing angle of the X-rays

path difference = $2d \sin \theta$

where d is the interplanar spacing. For constructive interference

$$2d \sin \theta = n\lambda \tag{9.3}$$

where $n = 0, 1, 2, 3, \ldots$ is the diffraction order. Equation (9.3) is called *Bragg's equation* and is extremely important in X-ray analysis.

Now simpler patterns than that derived from DNA can be obtained from regularly repeating structures that have a so-called one-dimensional symmetry, e.g. nerve myelin. Here the regularly repeating unit is derived from the layers of membranes that are wrapped round

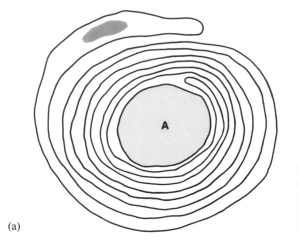

(a)

(b)

Fig. 9.5. Arrangement of Schwann cell membranes round the axon and the resulting diffraction pattern (SFD = 1.16 m).

the axon. The X-ray picture is composed simply of equally spaced lines (Fig. 9.5). Because of its simplicity, both visually and mathematically, nerve myelin will be the only system analysed in detail here.

Nerve myelin is formed from successive layers of Schwann cell membranes tightly packed together. A true repeating unit (as far as X-ray diffraction is concerned) consists of *two* membrane units together

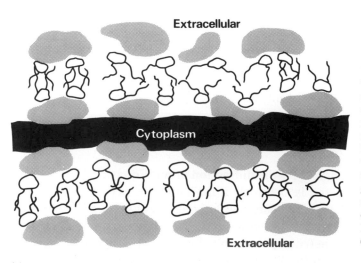

(a)

Fig. 9.6. (a) Diagram of a single repeat unit of Schwann cell membranes. Hatched areas represent membrane proteins. Osmium stain is taken up best by residual cytoplasmic material. (b) (*opposite*) Part of an amphibian nerve Schwann cell. The cytoplasmic regions can be seen as very dark lines and the extracellular regions, which take up less stain, are the faint lines in between. From the given magnification, the membrane repeat distance can be calculated and compared with the X-ray values from Fig. 9.12. (Magnification × 160 000.)

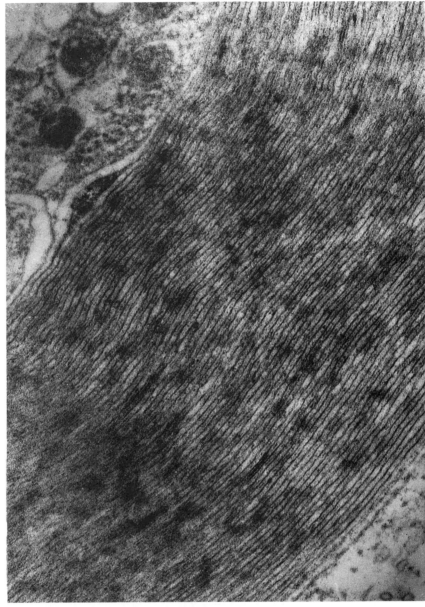

(b)

with a small amount of residual matter from former cytoplasmic and extracellular regions. The repeating unit can be seen in the electron microscope after staining with osmium tetroxide which is taken up particularly by the residual cytoplasmic proteins (Fig. 9.6b). These cytoplasmic regions are seen as dark lines in the electron micrograph. Sometimes visible between the dark lines are fainter lines representing the extracellular regions. We therefore have a quasi-crystalline structure in nerve myelin made up of repeating units of two membranes (Fig. 9.6a).

Description of a simple X-ray machine (low-angle)

All X-ray machines have three basic parts in common: (a) a *generator* to produce the X-rays; (b) a *monochromator* to provide radiation within a wavelength band; and (c) a '*camera*' to record the diffraction patterns.
(a)   In an X-ray generator, electrons are accelerated by high electric fields and strike a metal target (often copper) with high velocities. This produces a continuous spectrum with lines superimposed (Fig. 9.7). The accelerated electrons have sufficient energy to remove the copper electrons from their innermost orbitals (Fig. 9.2). When a *K*-shell electron is lost there is the possibility of an *L*-shell electron jumping down to take its place and such a transition gives rise to the so-called K_α line of copper (wavelength 0.154 nm). The K_β line is involved with the $M{\rightarrow}K$ transition and as the energy step is larger the wavelength involved (0.139 nm) is smaller.

Monochromatic X-rays are required for clearly defined diffraction patterns and in older machines this was crudely accomplished by removing the K_β line with a thin metallic filter. Nickel, for example, has an absorption edge at 0.148 nm (Fig. 9.7) so it will absorb the K_β line and allow K_α through. To the left of the absorption edge the photons have a high enough energy to eject electrons from the nickel *K*-shell,

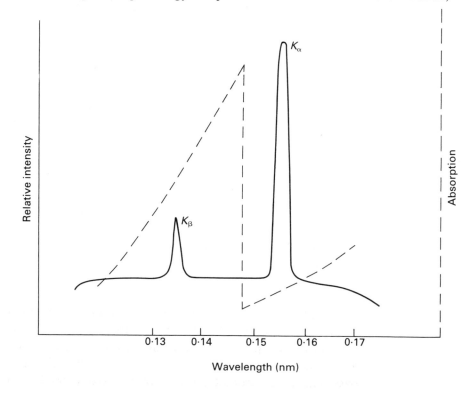

Fig. 9.7. Emission spectrum of copper target and the absorption spectrum of nickel.

Generator

Monochromator   Specimen holder   Film

Fig. 9.8. Ray diagram showing path of X-rays from generator to film.

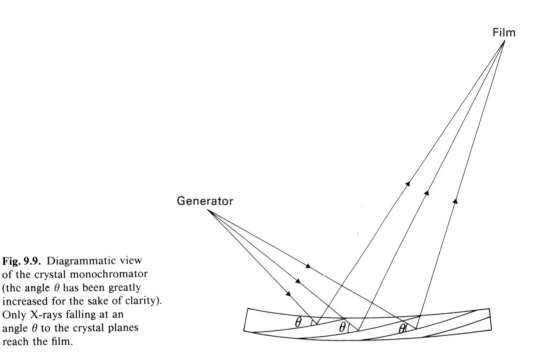

Film

Generator

Fig. 9.9. Diagrammatic view of the crystal monochromator (the angle θ has been greatly increased for the sake of clarity). Only X-rays falling at an angle θ to the crystal planes reach the film.

whereas to the right they do not have sufficient energy and the absorbance falls off. In modern machines filtering is achieved by means of a crystal monochromator rather than a nickel filter.

(b) The monochromator consists of a thin crystal which has been deformed to a curved shape so that the angle θ which the crystal planes make with the incoming rays is constant throughout the crystal (Figs 9.8 and 9.9).

A straightforward application of Bragg's equation shows why the X-rays are filtered

$$n\lambda = 2d \sin \theta \tag{9.4}$$

θ is set by fixing the crystal in position, and the interplanar spacing d is fixed by the nature of the crystal used so that X-rays of only one wavelength will pass through the system. By choosing the correct value of θ for the orientation of the crystal in the machine, only the K_α line will pass. The monochromatic X-rays are focused on the film because of the curved nature of the crystal.

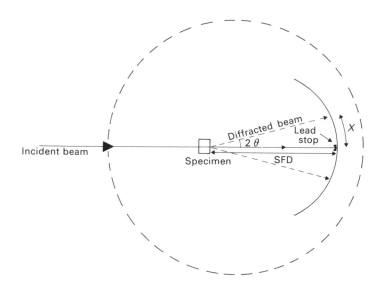

Fig. 9.10. X-rays enter the camera (a light-proof box shown in outline by dashes), are reflected from the crystal planes of the specimen and the diffraction pattern is recorded on film. A lead stop placed in front of the film prevents the very intense energy in zeroth-order diffraction beam from fogging a large area of film. SFD is the specimen to film distance and X is the distance of the first-order line from the straight-through position. Note that 2θ is the angle between the undiffracted and diffracted beams. The scattering angles involved are small, i.e. low angle.

(c) After leaving the monochromator the X-rays enter the camera which contains both the specimen to diffract the radiation and the film to record the pattern (Fig. 9.10). In more sophisticated machines the distribution of the diffracted energy is mapped out using radiation detectors (see Chapter 13) such as Geiger–Müller tubes or scintillation counters.

Analysis of diffraction pattern

Bragg's equation must again be invoked to understand the pattern. In this case λ and d are fixed so θ is prescribed from these. Hence only those planes of membranes that are oriented at the angle θ to the incoming beam give rise to a diffraction pattern (Fig. 9.11). When the X-rays diffracted at these planes constructively interfere, lines are observed corresponding to the first, second, third, etc. diffraction orders.

A photograph of a diffraction pattern from nerve myelin is shown in Fig. 9.12 and from this you can calculate the repeat distance d.

$$d = \frac{n\lambda}{2 \sin \theta}$$

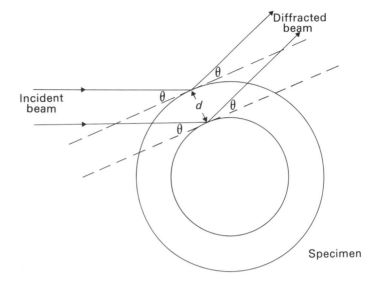

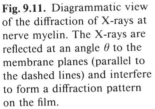

Fig. 9.11. Diagrammatic view of the diffraction of X-rays at nerve myelin. The X-rays are reflected at an angle θ to the membrane planes (parallel to the dashed lines) and interfere to form a diffraction pattern on the film.

Fig. 9.12. Low angle X-ray diffraction pattern from (a) fresh and (b) fixed amphibian nerve (SFD = 1.16 m). From the given specimen to film distance, the repeat distance for both patterns can be calculated and compared with the value from the electron microscope. In the case of a fixed nerve the lines are much fainter (because there is some disruption of structure) and their positions are marked by arrows.

and as the angles involved are small (low angle scattering)

$$d \approx \frac{n\lambda}{2\theta}$$

but

$$2\theta = \frac{X}{\text{SFD}} \text{ (Fig. 9.10)}$$

hence

$$d \approx \frac{n\lambda \text{ SFD}}{X}$$

where SFD is the specimen to film distance and when $n = 1$, X is the distance from the straight-through position to the first-order line.

In a simple one-dimensional repeating structure, e.g. myelin, the values for successive values of X should be in the ratio $1:2:3:4$ etc. In the fresh myelin pattern (Fig. 9.12a) you should find that they are in the ratio of $2:3:4$ indicating that the first diffraction line is hidden behind the stop. When calculating the various values of d, therefore, take $n = 2$ for the first measurable line, $n = 3$ for the second and so on. In this way a mean value for the interplanar spacing d is obtained. The mean value derived from the several diffraction orders can then be compared with the d value measured from the electron micrograph of known magnification. As the two values found may be quite different, a further pattern from a fixed nerve is provided (Fig. 9.12b) and some conclusions about the effects of fixatives on biological tissues can be made.

Problem 9.1

Careful examination of Fig. 9.12(a) reveals that the lines from myelin have alternating intensities — why does this occur? (Hint: What is the spacing of the diffraction lines from units that are spaced $d/2$ apart?)

9.4   Transmission electron microscope (TEM)

The electron microscope, in its various forms, is widely used for high resolution work in the investigation of cellular ultrastructure. It has very good resolution characteristics because of the small wavelengths associated with highly energetic electrons. This last statement might seem surprising but it has been shown experimentally that diffraction patterns are produced when electrons strike an ordered structure in much the same way as light waves do. Hence, not only do light waves have some of the characteristics of particulate matter (photoelectric

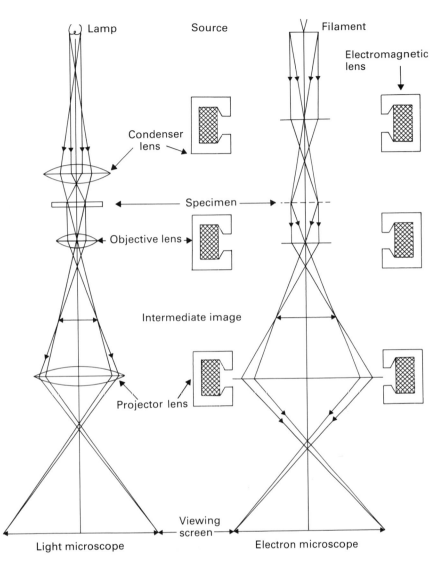

Fig. 9.13. Comparison of light and electron microscopes. In each instrument illumination from the source (lamp, filament in the electron gun) is focused by the condenser lens on to the specimen. A first magnified image is formed by the objective lens. This image is further magnified by the projector lens on to a ground glass screen (light) or fluorescent screen (electrons).

effect), but now we see that particles have waves characteristics.

De Broglie was the first to develop a theory to explain this and he derived the expression

$$\lambda = 0.1 \sqrt{\frac{150}{V}} \text{ nm} \tag{9.5}$$

where λ is the wavelength (nm) associated with an electron subjected to an accelerating voltage, V. In modern electron microscopes, as the electrons are accelerated by about 60 000 V, then $\lambda = 0.005$ nm and this is the theoretical limit of resolution.

In the transmission electron microscope (Fig. 9.13) electrons from the heated filament source are first focused on to the specimen by the condenser lens (as it is in the optical microscope). The first magnified image is formed by the objective lens, again in analogy with the optical microscope, and the final magnified image is projected on to a fluorescent screen or photographic film. Since electrons are scattered by air molecules the whole assembly, including the specimen, has to be housed under vacuum. The penetrating power of electrons is also very low so the specimen sections have to be extremely thin (10–50 nm). A cross-linking agent such as glutaraldehyde is generally used to 'fix' the specimen and it is embedded in a very hard resin before cutting with a diamond or glass knife.

The resolution of a modern transmission electron microscope is limited partly by specimen thickness and also by the spherical aberration of the instrument's objective lens. A point-to-point resolution of about 1 nm is the best that can be achieved at present. In the case of biological and inorganic specimens a further limitation is imposed by the radiation damage to the specimen caused by the electron beam. At normal illumination levels (more than 20 electrons s^{-1} 0.1 nm^{-2}) the resolution is limited to about 2.5–3.0 nm, whereas under minimum or low dose beam conditions the resolution can be improved to 1.5–2.0 nm.

Contrast in the TEM is derived from the scattering of electrons as they pass through the specimen. Since the angle of scatter increases with the atomic number of the specimen there is little contrast in biological materials as they are generally of low atomic number. For this reason heavy metal stains containing atoms of high atomic number (osmium tetroxide, uranyl salts, phosphotungstic acid, lead, etc.) have to be attached to biological specimens. A physical stop or aperture is normally placed below the specimen in the magnetic objective lens to increase contrast by preventing the scattered electrons from contributing to the final image.

The fluorescent screen on which the final image is projected is located at the base of the electron microscope and it can be further magnified optically with the aid of a binocular telescope. To record the image photographically a film camera is placed below the screen and the image is exposed on the film for a few seconds by raising the screen. The image can also be captured by a sensitive TV camera for computer-assisted image analysis.

The electrical and mechanical parts of the modern electron microscope have to meet very rigid conditions of operation as the vibrational movements of the entire system have to remain within the limits of resolution of the instrument (i.e. about 1 nm). Thus the electron microscope with its associated electronic and vibrational damping equipment has to be a complex and bulky piece of laboratory apparatus.

9.5 Special specimen preparation techniques

For hundreds of years replicas or casts have been used to preserve impressions of soft or transient objects. For example, Brewster made an isinglass replica of the lens in his diffraction studies (Chapter 1) and replica or casting methods are widely used for imaging biological specimens as they can be formed from high atomic number materials such as platinum or gold on a carbon support. The metal layer is often applied with the source at an angle to the surface of the specimen and a shadow casting is obtained, which gives additional information about the dimensions of particles (Fig. 9.14). This technique, combined with freeze-fracture, is extremely important in the study of membrane structure for example.

An important technique in the study of viruses and macromolecules is negative staining where the specimen is embedded in a droplet of an electron-dense material such as uranyl acetate or phosphotungstate. These stains penetrate the empty spaces between the macromolecules which appear light against a dark background. Contrast can be further enchanced by image-processing techniques.

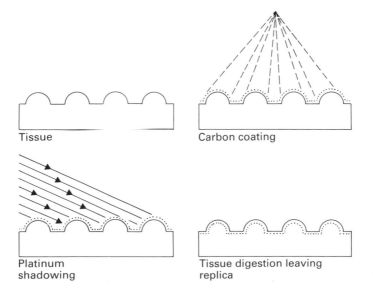

Fig. 9.14. The tissue in the fixed or frozen state is first coated with a thin layer of carbon and then a thin layer of an electron-dense material such as platinum is applied. The tissue is then digested away leaving a very thin replica of the surface details of the original tissue.

9.6   Scanning electron microscope

Very beautiful three-dimensional views of the surface structure of biological materials can be obtained with the scanning electron microscope (SEM) and the same basic system for producing and focusing electrons is used in both the SEM and TEM. However, in the SEM, the electron optical system is designed to form a small beam of electrons to scan the surface of the specimen and the secondary electrons are collected by a detector (Fig. 9.15). The resolution of the SEM is limited by the final diameter of the electron probe (about 2.5–3.0 nm).

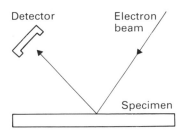

Fig. 9.15. Diagram of mode of operation of the scanning electron microscope (SEM). The narrow electron beam scans across the tissue and the intensity of the secondary electrons, recorded at the detector, is displayed as a two-dimensional image on a television screen.

The electron beam is arranged to scan the specimen by using magnetic scanning coils at scanning and frame rates similar to television systems. The image is derived from the production of secondary electrons emitted by the specimen when scanned by the probe. It follows that biological specimens have to be coated with some layer of heavy metal or coated with a conducting material to increase secondary emission and prevent the object charging up.

Unlike the TEM the specimens capable of being studied in the SEM can be physically large, and a suitable chamber is fitted to the base of the microscope. Large pieces of tissue or intact biological structures of several centimetres across can be examined in the SEM. The mechanical stage allows translation across the area of the scanning probe and also permits the specimen to be rotated and tilted through relatively wide angles. The final image obtained with the SEM is displayed on a cathode ray tube (CRT) with a scan and frame interval synchronized with the SEM system. Photographs are recorded by placing a camera in front of the CRT screen. Contrast in the image can be adjusted electronically.

Electrons in the probe beam also have sufficient energy to induce X-ray emission from elements within the tissue. As each element has its characteristic X-ray emission spectrum (Fig. 9.7), the SEM can be used to obtain information about the composition of the specimen by analysing the energy spectrum of the emitted X-rays (Fig. 9.16). There are at present two major techniques for performing the analysis: (a)

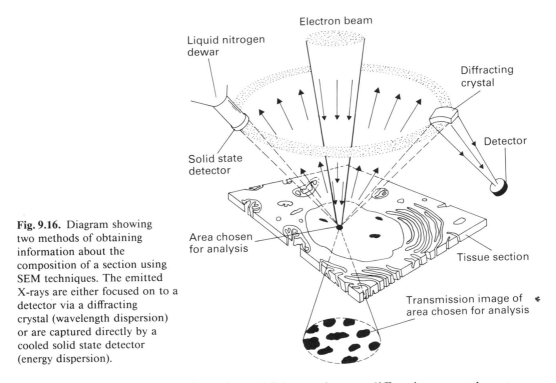

Electron beam

Liquid nitrogen
dewar

Diffracting
crystal

Detector

Solid state
detector

Area chosen
for analysis

Tissue section

Transmission image of
area chosen for analysis

Fig. 9.16. Diagram showing two methods of obtaining information about the composition of a section using SEM techniques. The emitted X-rays are either focused on to a detector via a diffracting crystal (wavelength dispersion) or are captured directly by a cooled solid state detector (energy dispersion).

wavelength dispersion which requires a diffracting crystal system (Fig. 9.17), and (b) energy dispersion, where the whole X-ray energy spectrum is first captured by a solid state detector and is then processed electronically (Fig. 9.18).

9.7 Tissue elemental analyses using EM techniques

Wavelength dispersion

The wavelength dispersive spectrometer consists of a curved crystal monochromator (as in Fig. 9.9) and radiation detector and analyser system. In a fully-focusing spectrometer, the specimen, crystal and detector lie on the same circle. The X-rays emitted by a small volume of the sample are analysed by the monochromator and if this is set at the Bragg angle for a given wavelength then only X-rays of that wavelength are reflected and focused on the detector window. Their intensity is then measured after discrimination by a pulse height analyser (Fig. 9.17).

Energy dispersion (EDXA)

The energy dispersive spectrometer consists of a solid state semi-conductor radiation detector (often silicon impregnated with lithium)

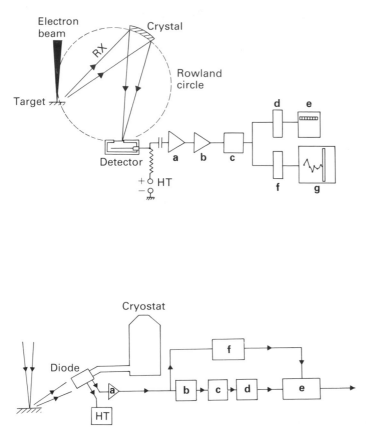

Fig. 9.17. Diagram of a wavelength dispersive X-ray spectrometer: (a) preamplifier; (b) amplifier; (c) single channel analyser; (d) timer; (e) scaler; (f) integrator; (g) recorder.

Fig. 9.18. Diagram of an energy dispersive X-ray spectrometer: (a) preamplifier; (b) amplifier; (c) base-line restorer; (d) analogue to digital (A/D) converter; (e) multichannel analyser; (f) pulse pile-up rejection on some instruments. The solid-state detector (diode) is cooled to liquid nitrogen temperatures to reduce background (electronic) noise (after Maurice *et al.*, 1980).

which collects the entire X-ray spectrum and transfers it to a multichannel analyser which stores the different spectral lines as a function of their energy (Fig. 9.18). The solid-state detector produces voltage pulses that are proportional to the energy of the impinging X-rays. A field-effect transistor (FET) system is used to amplify the voltage pulses and, in order to reduce thermal noise to a minimum, the detector assembly is housed in a liquid nitrogen cryostat system.

Elemental analyses of biological tissues

In the past 10 years sample preparation and analytical techniques have advanced at an equally rapid rate so that it is now possible to carry out elemental analysis of thin sections of biological tissues that have been snap frozen. No fixative is required so there is no leaching or redistribution of tissue elements. Greater sensitivity is achieved if the water is removed from the sample and therefore many tissues are analysed after critical-point freeze-drying (again to prevent redistribution of constituent elements). The tissue is often fractured and analysed by SEM techniques. Figs. 9.19, 9.20 and 9.21 show several techniques

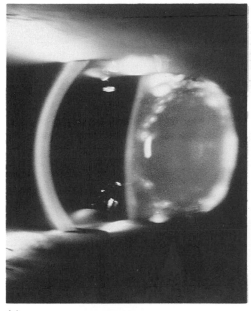

(a)

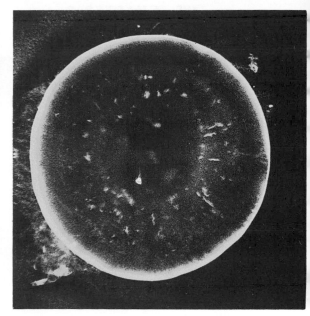

(b)

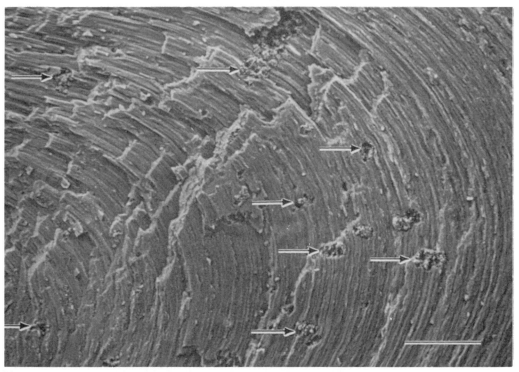

(c)

Fig. 9.19. (a) Slit lamp camera photograph of a human cataractous lens with small localized opacities. (b) Lens with localized opacities photographed against a dark background after surgery. (c) Low-magnification scanning electron micrograph of the lens shown in (b) after fixing (glutaraldehyde), critical-point drying and fracturing. Arrows indicate distributions of localized areas of degeneration corresponding in size and distribution to the opacities seen in the photograph. In subsequent photographs, these are referred to as 'opacities' (gold-coated; bar = 100 μm). (d) (*overleaf*) Details of an individual punctate opacity and surrounding normal fibre cells, which show characteristic interdigitations (gold-coated; bar = 20 μm).

Fig. 9.19
(*continued*) (d)

for the analysis of a cataractous lens from its appearance in the patient (Fig. 9.19a) to a detailed ultrastructural investigation of the form and composition of the light-scattering regions (Harding *et al.*, 1982).

Figure 9.20 shows the energy spectrum obtained on analysing different regions within a cataractous lens. Very high phosphorus values are obtained at the centre of the light-scattering body. The total surface of the freeze-fracture fractured face of a sample can be scanned with the energy-dispersive spectrometer set sequentially at different energy values (corresponding to phosphorus, sulphur, etc.). The two-dimensional distribution data can be stored digitally and then plotted alongside the conventional SEM image (Fig. 9.21a and b).

The lens proteins (crystallins) are rich in sulphur (Fig. 9.20b), while the composition of the light-scattering region is quite different (Fig. 9.20a). The relatively high phosphorus peak is characteristic of membranes and a TEM micrograph of this region shows a parallel stack of them (Fig. 9.21d). The refractive index of this region would be

Fig. 9.20. (*opposite*) (a) EDXA spectrum obtained by focusing electrons on a small region of a punctate opacity (open arrow in Fig. 9.21a). The spectrum shows a strong peak at the phosphorus energy level (P), and a small peak for sulphur (S). (Accelerating voltage = 10 kV.) (b) EDXA spectrum of a small spot on a normal fibre cell (solid arrow in 9.21a) adjacent to punctate deposit. The spectrum shows a strong peak at the sulphur energy level (S), and a small peak for phosphorus (P). (Accelerating voltage = 10 kV). (c) EDXA scan from a normal human lens obtained with a high resolution system. Although only two peaks are identified in the above study corresponding to sulphur and phosphorus, the human lens energy dispersion spectrum consists of many more elements and a more detailed spectrum for a normal human lens has identifiable additional peaks corresponding to Na^+, Cl^-, K^+ and Ca^{2+}. The latter peak increases dramatically in many types of cataract (x axis: X-ray energy; y axis: X-ray intensity (cts s^{-1})).

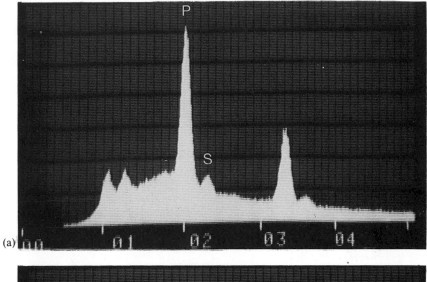

(a)

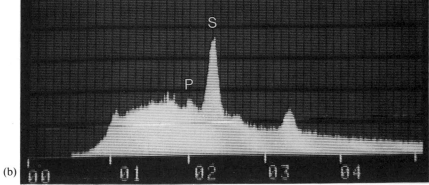

(b)

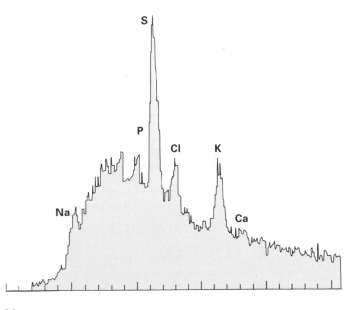

(c)

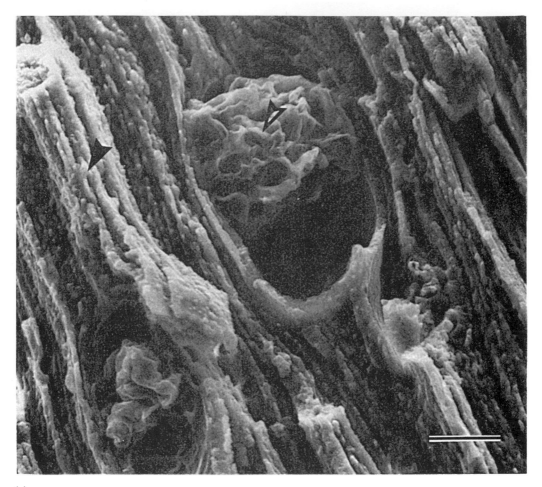

(a)

Fig. 9.21. (a) Area from the lens shown in Fig. 9.19c, refractured to remove the gold coating and recoated with carbon in preparation for EDXA. Two punctate deposits are evident. The two arrows indicate spots at which analyses for phosphorus and sulphur were made (see Fig. 9.20 a,b). Open arrow: material within the opacity; solid arrow, normal lens fibre cells. (Carbon coated; bar = 10 μm.) (*opposite*)
(b) EDXA mapping of phosphorus for same field shown in (a). The two high phosphorus areas have light regions the same shapes as the two punctate opacities.
(c) Transmission electron micrograph of single opacity, showing almost parallel arrangement of membranes. (c) and (d) are from glutaraldehyde-fixed and critical point-dried material that was used for SEM and later postfixed in osmium, embedded, and sectioned for TEM. (Bar = 0.5μm.) (d) Higher magnification transmission electron micrograph of portion of an opacity. Unit membrane structure is evident here, as it is in all other punctate opacities examined with TEM. (Bar = 0.1 μm.) (Original photographs for Figs 9.19, 9.20 and 9.21 kindly provided by Dr C.V. Harding and used with permission.)

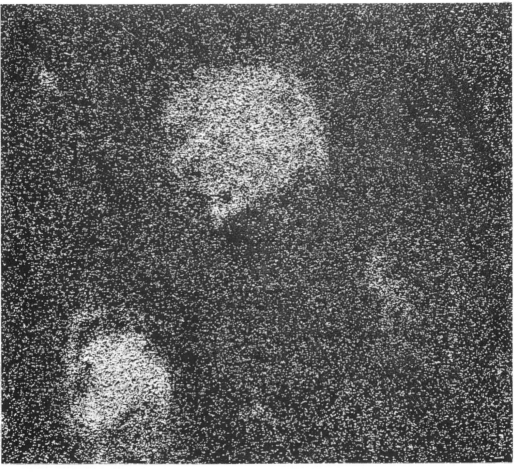

(b)

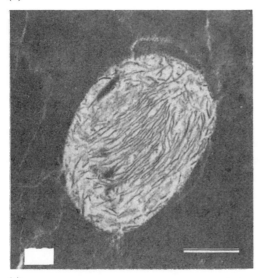

(c)

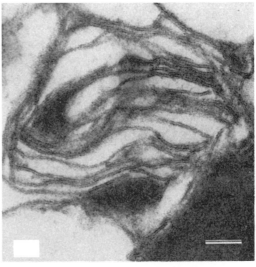

(d)

expected to be quite different from that of the surrounding fibre cells and this explains the increase in light scatter from these regions. The mechanism responsible for the appearance of regularly stacked membranes is not understood, but it is interesting that such figures occur in more than one type of cataract.

9.8 Image analysis

Sir David Brewster was probably the first to use diffraction techniques to obtain information about regularly repeating structures within a biological specimen. He had to use somewhat complex mathematical techniques developed by Fourier to interpret the patterns he measured by eye (Chapter 1). In analyses being undertaken at present, Brewster's candle has been replaced by the laser, his eye by a battery of photocells and his mathematically-driven quill by the computer. In place of cod lens fibres are to be found molecular arrays in viruses, bacterial cell walls, mammalian cell membranes and, at the other extremity, naked atoms within thin foils. Techniques are also being developed to study untreated biological assemblies by fast-freezing methods.

Image processing and analysing

Highly regular images provide the best subjects for image processing as the image can be improved either by photographic superposition methods or computer analysis. For example, in the case of images with rotational symmetry, the image is projected on to photographic paper at a convenient enlargement. If it is suspected that the image has a sixfold symmetry, for example, then the photographic paper is rotated about the centre of the particle. An exposure is made after each rotation of 60° for one-sixth of the exposure time required to make a normal photographic print of the particle. The background noise does not have rotational symmetry and is reduced at each rotation. If n rotations are made, then the background noise is reduced by a factor $\sqrt{n}$.

In computer-enchancing methods, the image of the particle is transformed from Cartesian (x,y) to polar (r,θ) coordinates with the origin of the polar coordinates corresponding to the rotational centre of the particle. With this coordinate system, the particle density exhibits periodicity in the angle θ and a Fourier analysis can be carried out to determine the periodic elements. The non-periodic elements (noise) can then be subtracted from the total to give a final filtered image.

Thin gold films are excellent subjects for study in the TEM. The lattice is regular and as the interatomic spacing (0.209 nm) is within the limit of resolution of the electron microscope, such a foil can be used to

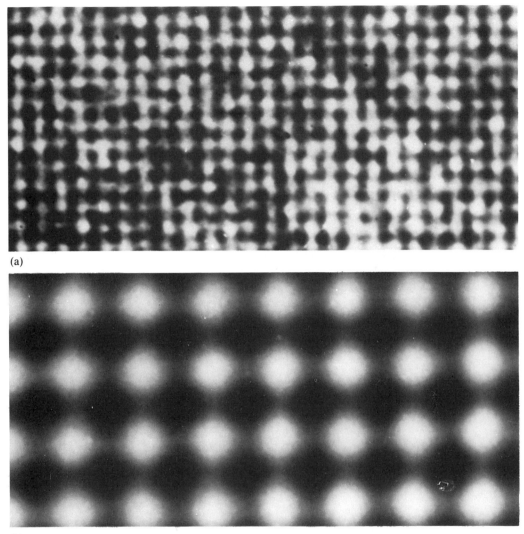

(a)

(b)

Fig. 9.22. (a) Gold lattice image obtained from thin film specimen obtained by TEM. (b) Enhanced image obtained by photographic superposition. Centre to centre spacing = 0.209 nm (from Beeston *et al.*, 1972).

calibrate photographs and images obtained with the microscope. The image obtained with a gold film, even in a good TEM, can be greatly improved by image processing techniques.

Averaging techniques

Figure 9.22(a) shows the regularly repeating, square array of gold atoms in the lattice, but a significant amount of 'background noise' is present arising from within the electron microscope and from the photographic processing. The image can be greatly enhanced (Fig. 9.22b) by making several exposures of the negative of Fig. 9.22(a) on photographic paper and translating the paper through a distance (either horizontally or vertically) equivalent to 0.21 nm each time an exposure is made. The image could also be reconstructed from the optical diffraction pattern from a gold foil photographic negative.

Optical diffraction techniques

The basic principle of using visible light to form diffraction patterns is illustrated in Fig. 9.23(a). Coherent light is provided by a laser (Section 9.10) at A and a diffraction lens is shown at B. A periodic object such as an electron micrograph negative is placed at C, forming a diffraction

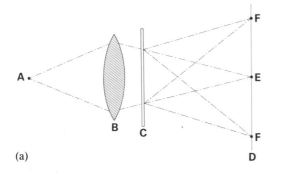

(a)

(b)

Fig. 9.23. (a) Method for obtaining optical diffraction patterns from electron micrograph photographic negatives. (b) Optical transform of gold lattice in Fig. 9.22(a). The bright first-order spots correspond to a spacing of 0.209 nm (from Beeston *et al.*, 1972).

pattern (or optical Fourier transform) in the diffraction plane D. The extent of the diffraction pattern (at D, E and F) will depend on the level of ordering, distribution, area and resolution of the detail in C and also on the signal-to-noise ratio in the original image.

Since the gold-foil electron microscope image consists of a pattern of regularly repeating dots, it forms an excellent diffraction pattern when placed in position C (Fig. 9.23b). The pattern is characteristic of a square array of scattering centres and the diffraction spots correspond to a lattice spacing of 0.209 nm. The 'background noise' can be seen to be partly random in character (faint discrete spots) and partly regular in the form of rays radiating from the centre. Both can be filtered, or removed, and the remaining spots can be used to reconstruct the original image, by computer-assisted Fourier analysis. It would in fact, look very similar to the averaged image in Fig. 9.22(b). Some virus particles show remarkable regularity of substructure and Fig. 9.24(a) shows the lattice array from cowpea chlorotic mottle virus (CCMV) prepared by negative staining techniques (TEM micrograph). There is some evidence of sub-structure at the periphery of each virus particle and of bridge structures and these become much clearer after photographic averaging (Fig. 9.24b). The optical diffraction pattern (Fig. 9.24c) obtained from a negative of Fig. 9.24a and its computer-generated Fourier transform (Fig. 9.24d) show clear evidence of a twofold symmetry in the pattern. This, and other more detailed analysis, permitted a computer-reconstructed image to be made (Fig. 9.24e). This compares well with the photographically averaged image of the virus array obtained by diagonal superimposition of the original image (Fig. 9.24b). Each virus particle is viewed along a twofold rotational symmetry axis but neighbouring particles are rotated by 90° in the lattice. Finally, all of the assembled information can be used to construct a three-dimensional scale model (Fig. 9.24f). Such a model is now usually constructed digitally using three-dimensional imaging computer software (Horne, 1979).

9.9   Absorption of light and fluorescence

The photoelectric effect (Fig. 9.1) demonstrated the dual nature of electromagnetic radiations. Under some circumstances, e.g. in the phenomena of light diffraction and interference, the resulting patterns can most readily be explained in terms of the wave theory. In other circumstances, when light interacts with matter and is then absorbed, and sometimes re-emitted, the only explanation is in terms of quantum theory. As biologists we are concerned mainly in the interactions of light with polyatomic molecules rather than simple atoms. Quantum theory states that electrons in molecules are arranged in energy levels just as

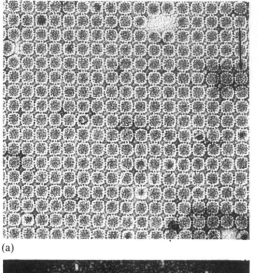

(a)

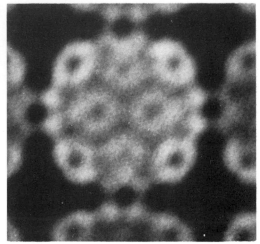

(b)

(c)

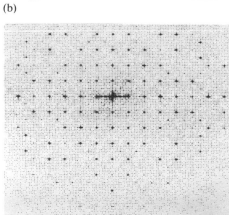

(d)

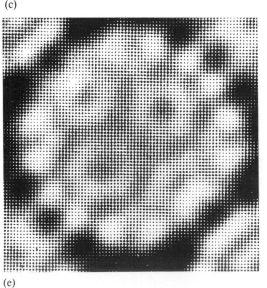

(e)

(f)

Fig. 9.24. (*opposite*) (a) Lattice array from cowpea chlorotic mottle virus prepared by negative staining methods. The square array provides little evidence of packing orientation of neighbouring particles within the array. There is some evidence, however, of subunit structure at the periphery of each virus particle.
(b) Photographically averaged image of CCMV lattice array by diagonal superimposition of the image. (c) Optical diffraction pattern of the square array of CCMV. The first-order spots correspond to a spacing of 27 nm. (d) Computer-generated Fourier transformation of the optical diffraction pattern (c). (e) Computer-reconstructed image using the information obtained in (a–d). (f) Scale model of CCMV built from information obtained in (a–d). (Photographs kindly provided by Professor R.W. Horne and used with permission.)

they are in atoms, the only complication being that there are many more possible energy levels for the electrons (Fig. 9.25).

When light impinges on matter, the probability of it being absorbed depends on the energy of the photon and the distribution of energy levels in the matter. The photon only has a significant probability of being absorbed if there are two energy levels whose energy difference exactly matches the energy of the photon.

In some substances, e.g. glass, the difference between energy levels is so great that the relatively low energy visible light photons are not absorbed and the material is transparent. Good conductors of electricity, on the other hand, have electrons that are free to move into a vast range of energy states so that almost every incident photon is likely to be absorbed. However, in most cases the photons are rapidly re-emitted in a backward direction and as a result thin metal sheets appear quite opaque as all the photons are reflected.

When light falls on a suspension of biological molecules, a certain fraction of the light is absorbed, and the amount absorbed depends both on the molecules comprising the suspension and the wavelength of the

Fig. 9.25. Comparison of electronic energy states in atoms and molecules. G is the ground state and E_1 and E_2 are two possible excited states. In molecules the total number of energy states is increased because of molecular rotations and vibrations.

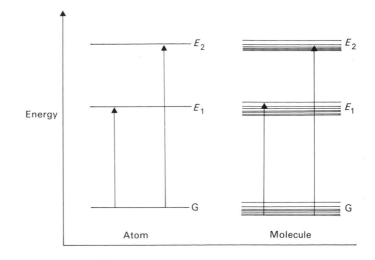

light. When the absorption versus wavelength characteristics are analysed with the aid of a *spectrophotometer* several absorption bands are often evident. The reason why there are bands rather than lines lies in the fact that polyatomic molecules can no longer be treated as small rigid particles. The rotation of the molecule and the vibrations of its constituent nuclei must be taken into account. The energies of these motions are also quantized and the resulting energy levels must be added to the atomic energy levels to obtain a complete picture of the available energy states.

A picture of how the vibration levels may be built up can be obtained most simply from the consideration of a diatomic molecule. In the lowest energy state the amplitude of vibration of the nuclei is small and as energy is pumped into the molecule, e.g. by heating, the amplitude of vibration will increase and other energy levels will be created. In this context quantum theory states that only certain amplitudes of vibration are allowed and so discrete energy levels are created. A graph of potential energy against internuclear distance has a roughly parabolic form (curve G in Fig. 9.26a), and possible vibrational energy levels are

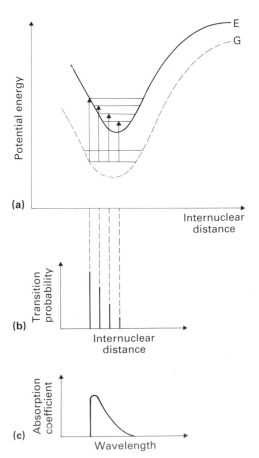

Fig. 9.26. (a) Diagrammatic representation of the vibrational energy levels in a diatomic molecule. G and E are the ground and excited state respectively. (b) Four possible electronic transitions are shown and their transition probablities are given. As nuclei spend most of their time at the extreme position of the vibration, the furthest left transition is the most likely. (c) An absorption band is finally observed because rotational energy levels broaden the individual lines. (after Thomas, 1965.)

shown for the electronic ground state of a diatomic molecule by the horizontal lines. In the same manner a curve can be drawn for an electronically excited molecule E. When photons are absorbed, electrons will jump from a ground state energy level to an excited state level. Let us suppose for simplicity that only the lowest energy level is occupied in the ground state. The very important Franck-Condon principle states that as electrons can move faster than nuclei, the latter do not have time to move during a transition, and so the absorption act (which takes approximately 10^{-8} s) can be represented by a vertical arrow. As nuclei spend most of their time at the extreme position of their vibration, the most probable transition is represented by the arrow furthest to the left. The other transitions are less likely. The probability of a transition from the ground state, i.e. the probability that a photon will be absorbed, therefore depends on the internuclear separation (Fig. 9.26b) and as this separation determines the energy of the molecule, the transition probability will also depend on the energy of the photon (and hence on its wavelength). Because the molecules have rotational as well as vibrational energy the individual lines are broadened to give an *absorption band* (Fig. 9.26c). Here we have only dealt with transitions from the lowest vibrational energy level of the ground state at room temperature, but the next level above this should also be considered. The complete absorption spectrum will therefore be more complex than that shown in Fig. 9.26(c). It will in fact also have minor absorption bands on the short wavelength side of the absorption maximum.

Although the true state of affairs is much more complex in polyatomic molecules, absorption takes place along much the same lines. When a molecule has absorbed a photon it is in an excited state and has an excess of energy which it must be rid of — there are several means of achieving this. In the transitions $G \rightarrow E_2$ and $G \rightarrow E_3$ (Fig. 9.27) the electrons return within 10^{-12} s to the excited state E_1 and in these

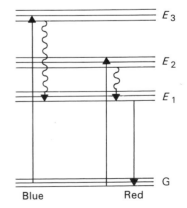

Fig. 9.27. The energy levels of a polyatomic molecule. Part of the energy absorbed in the transition $G \rightarrow E_3$ is lost due to radiationless transfer through overlapping levels to E_1. The energy content of the photon emitted in the transition $E_1 \rightarrow G$ is less than that absorbed and hence the wavelength is shifted from blue $\rightarrow$ red.

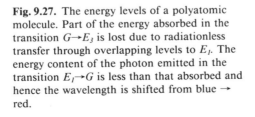

transitions energy is lost in the form of heat. In some molecules the electrons remain in E_1 for over 10^{-8} s and when they eventually return to the ground state they emit light; in fact they *fluoresce*. This internal arrangement of electrons can be seen in chlorophyll which, though it absorbs in the blue and in the red, only fluoresces in the red.

Electrons can become trapped in excited energy levels called triplet states, from which they cannot easily drop to the ground state. When the electron from an excited triplet state drops to the ground state, the light emitted is called *delayed fluorescence* if it takes place within 10^{-3} s, otherwise it is known as *phosphorescence*. Molecules which possess these triplets states are highly reactive and the excitation energy can be passed on to other molecules if they have matching energy levels. Chemical energy e.g. in the form of dissociation energy, can be passed on in this way.

9.10   The laser

Light emission takes place when an excited atom loses energy on its return to the ground state. The transitions occur over an extremely short period of time (10^{-9} s) and so the light from a conventional filament or gas vapour source changes in phase and plane of polarization every 10^{-9} s. Such sources are said to be non-coherent. Very powerful *coherent* sources can be obtained by trapping electrons in a long-lived metastable energy state. From there they can be stimulated to make a transition to the ground state by photons of the correct energy.

Excited electrons can return to the ground state by two mechanisms: either spontaneously, emitting a photon in the process, or by stimulation if they collide with a photon of the same energy as they emit. In normal substances most of the electrons are in the ground state so when light impinges on such a substance there is very little stimulated emission as most of the incoming energy is used to promote electrons from the ground to the excited state. In order that amplification can occur it is first necessary to produce an electronic population inversion so that more electrons are in the excited than in the ground state. In the ruby laser (Fig. 9.28a), the initial excitation is obtained from a xenon flash tube which is wrapped round a ruby rod. A 500 μs flash of xenon light excites the ruby molecules to an energy level (absorption band) where they only remain for 5×10^{-8} s (Fig. 9.28b). From there they make spontaneous transitions preferentially to a metastable level where they remain for as long as 3×10^{-3} s. Hence, with one intense exciting flash a population inversion has been achieved.

When a molecule finally makes a transition from the metastable to ground state, it emits a photon with wavelength 694.3 nm. This photon can now interact with an excited molecule in a metastable state when a

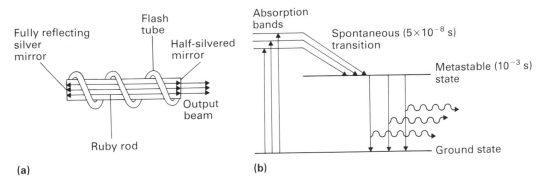

Fig. 9.28. (a) Diagram of a ruby laser. A xenon tube is wrapped around a ruby rod which has a fully-reflecting silver mirror at one end and a half-silvered mirror at the other. The output consists of an intense, parallel beam of monochromatic (694.3 nm), coherent light. (b) Electronic energy level transitions within the ruby rod leading to the emission of an intense beam of photons of one wavelength.

second photon is emitted. This second photon can now collide with another metastable molecule and in this way an avalanche of stimulated light emission is produced. The efficiency of the laser is considerably increased by having a fully reflecting surface at one end of the rod and a half-silvered mirror at the other end. The light photons produced therefore reflect internally up and down the rod, greatly increasing the probability of interaction and the metastable energy is released in a single red flash of wavelength 694.3 nm at the half-silvered end.

In the laser, therefore, an intense parallel beam of monochromatic, coherent light is produced and light with these properties has a very wide range of uses. The laser is used in eye surgery, for example, in repairing retinal detachments where the light is focused by the cornea and lens on to the tear at the back of the eye. It is used in the investigation of protein diffusion, both in the fluorescence recovery after photo-bleaching technique (FRAP) where the very high energy content of the light is again important and in the quasi-elastic light-scattering technique where the coherence properties are used. We have seen previously that the monochromatic nature of the light is used in image analysis methods (Section 9.8).

9.11 Photosynthesis: energy transfer and fluorescence

The coupling between carotenoids and chlorophyll is an example of energy transfer from molecule to molecule. That two pigment systems are involved in photosynthesis is shown by the fact that photosynthesis effectiveness, e.g. measured in terms of CO_2 fixation, does not fall to very low values in the region between the two chlorophyll absorption peaks (Fig. 9.29). It was therefore proposed that the energy absorbed by

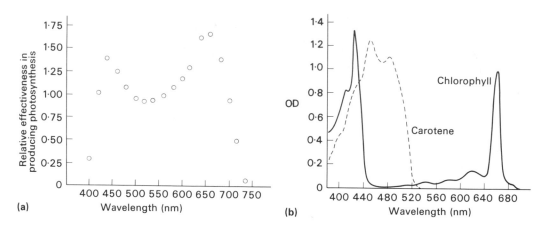

Fig. 9.29. (a) The effectiveness of various wavelengths of light in producing photosynthetic carbon dioxide fixation. Two clear peaks are in evidence. The fact that the effectiveness does not go down to zero between the peaks indicates the existence of at least one more peak in the wavelengths between the two (Epstein, 1963). (b) The absorption spectra of chlorophyll and carotene. The chlorphyll peaks at about 430 and 655 nm and the carotene peaks near 450 and 490 nm are identified as being responsible for the action spectrum peaks in (a). (Epstein, 1963, reproduced by permission of Addison-Wesley Inc.)

carotene can be passed on to the chlorophyll. This hypothesis is further backed up by spectroscopic evidence which shows that when organisms are irradiated with 500 nm light (absorbed only by the carotenoids) there is no carotenoid fluorescence, but instead there is the typical red fluorescence from chlorophyll.

Chloroplast fluorescence *in vivo* is in fact normally weak as much of the energy absorbed by the chlorophyll–carotene system is utilized in a series of reactions that produce finally ATP. Weedkillers are generally chemicals that block one or more of these reactions and, as a result, chloroplast fluorescence is considerably enhanced, while ATP production is reduced.

9.12   Fluorescence microscopy

The fluorescence microscope is now an essential piece of equipment in most hospital and biological laboratories because of the recent development of immunofluorescent antibodies. These allow specific molecules (antigens) to be detected within a normal or cancerous cell and the role of these molecules in cell function can be studied.

Suppose that we are interested in investigating the role of tubulin, a cytoskeletal or structural protein, in the growth of cancerous cells; one way of doing this is to prepare antibodies by injecting rabbits with tubulin. A fluorescent marker (e.g. fluorescein isothiocyanate, or FITC) is chemically attached to the antibodies and, when a microscope slide

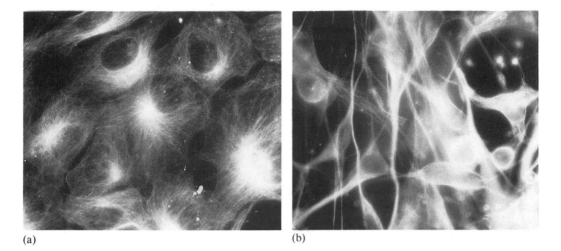

(a)   (b)

Fig. 9.30. (a) Photograph of FITC-labelled tubulin in the fluorescence microscope. (a) Normal tissue-cultured cells where the tubulin radiates out from a specific area in the cell. (b) Tubulin in transformed cells where shape of tubulin organization has changed. (Photographs kindly provided by Drs A.R. Prescott and R.M. Warn.)

with a tubulin-containing cell is exposed to an antibody solution, the antibodies specifically react with tubulin in the cell. After thoroughly washing the cells, the labelled tubulin which remains can be clearly seen in the microscope (Fig. 9.30). Both direct and indirect labelling techniques can be used (Fig. 9.31).

The main elements of the fluorescent microscope consist of a high intensity light source, excitation filter, specimen, barrier filter and recording device (Fig. 9.32). The absorption and fluorescence spectra of FITC are shown in Fig. 9.33(a) and from their characteristics it is possible to choose a convenient set of filters. The transmission characteristics of one possible set are given in Fig. 9.33(b). The excitation filter consists of a band pass filter (450–490 nm) while the barrier filter is a longwave pass filter, or cut-off filter (>500 nm).

Modern fluorescent microscopes do not operate in the transmitted light mode but use reflected light or *epifluorescence*. In this type of microscope the excitation light, after passing through the first filter, is reflected through 90° at an interference mirror. This is a specially constructed mirror which reflects all light of wavelengths less than 510 nm on the specimen (FT 510 nm in Fig. 9.34a). Light of longer wavelengths passes straight through the mirror. The short wavelength light is then focused on to the specimen by the condenser lens (Fig. 9.34b). The fluorescent light of longer wavelength, emitted by the specimen, passes through the same lens which now acts as an objective lens. This long wavelength light is transmitted through the beam splitter, while shorter wavelengths are reflected. Before the final image

Normal antigen–antibody reaction
where AB is, for example,
a human antibody against
the antigen AG

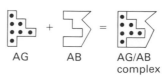

AG   AB   AG/AB
complex

'Labelling'

AB Fluorochrome Fluorochromized AB

Immunofluorescence (direct).
Labelled antibody reacting with antigen
= fluorescence after excitation

AG   AB (fluor.) AG/AB-complex (fluor.)

Antibodies can only react with
specific antigens

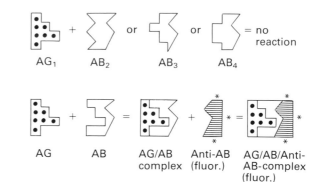

AG_1 AB_2 AB_3 AB_4

Indirect method: specific antibody
reacts with antigen; labelled anti-
antibody reacts with (all human)
antibodies = fluorescence after first
reaction

AG   AB   AG/AB   Anti-AB   AG/AB/Anti-
complex (fluor.)   AB-complex
(fluor.)

Fig. 9.31. Diagram of immunofluorescence reactions. The direct method is
described in the text, but the more common indirect method avoids storing specifically
labelled antibodies for each antigen. An intermediate host such as rabbit is injected
with *human* antibodies and such animals immediately produce antibodies-against-
human-antibodies (Anti-AB). These antibodies can be fluorescently labelled and will
react with every human antibody complex as if it were an antigen.

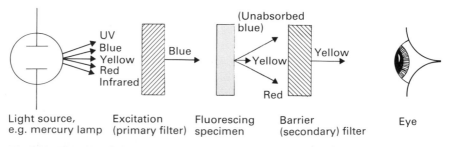

Light source,   Excitation   Fluorescing   Barrier   Eye
e.g. mercury lamp (primary filter) specimen (secondary) filter

Fig 9.32. Diagram of elements in a fluorescence microscope for blue excitation
(FITC method, for example). The lamp emits a range of wavelengths but the primary
filter only allows through the blue wavelengths that excite FITC molecules. These
fluorescence in the yellow and red and the unabsorbed blue and also red wavelengths
are blocked at the second narrow-band filter. Only the yellow wavelengths reach the
eye (after Holz, 1977).

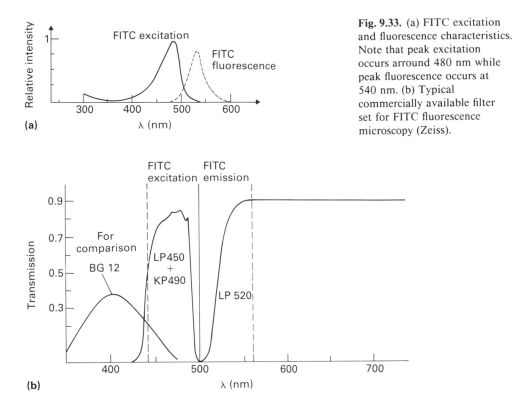

Fig. 9.33. (a) FITC excitation and fluorescence characteristics. Note that peak excitation occurs arround 480 nm while peak fluorescence occurs at 540 nm. (b) Typical commercially available filter set for FITC fluorescence microscopy (Zeiss).

is formed the light also passes through a long wavelength pass filter (LP 520 in Fig. 9.33b). Figure 9.30 shows images of tubulin labelled with FITC antibodies which were obtained from a Zeiss epifluorescence microscope.

Use of fluorescence microscope in measuring cell calcium and pH

Recently certain fluorescent dyes have been developed where the fluorescence spectrum is sensitive either to the ambient calcium or proton concentration (Fig. 9.35). The value of these dyes for cellular work lies in the fact that they have a hydrophobic group attached through an ester link which renders them lipid soluble. When added to the external solution the dye readily diffuses through the plasma membrane into the cell where the hydrophobic group is removed by cytoplasmic esterases and the dye is 'trapped'. The measurement is usually made by irradiating the cell at two wavelengths in succession while monitoring the fluorescence at a single wavelength. If one of the wavelengths chosen is the isosbestic point (360 nm), a qualitative measure of internal calcium can be obtained from the ratio of the

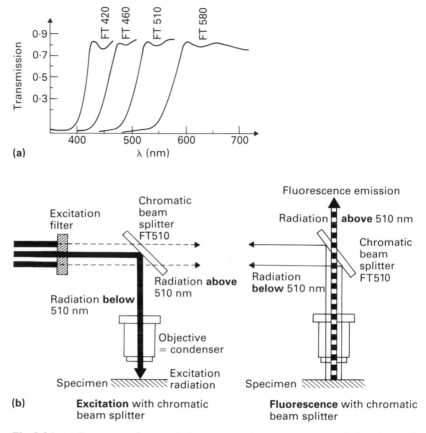

Fig. 9.34. (a) Examples of transmission curves of commercially available chromatic beam splitters. The *reflection* curves for each are the mirror-inverted images of the above. (b) Epifluorescence system for FITC-labelled specimens (after Holz, 1977).

fluorescence at the two exciting wavelengths. This technique can also be combined with image analysis methods to give the spatial distribution of calcium or pH within a cell. The concentration estimate can be obtained in absolute terms by equilibrating the cell cytoplasm with the external solution at the end of the experiment. The cell membrane is usually permeabilized for this measurement by adding the calcium ionophone, ionomycin, which allows the calcium to enter but not the dye to leave.

A recent example of the use of this technique to investigate the role of calcium in cell division is shown in Figs 9.36 and 9.37. When an antibody to the calcium sequestring protein (46 K protein) is injected into one cell of a two-cell embryo, the calcium concentration increases from 0.2 to 1.0 μM and the mitotic spindle (monitored by polarizing optics) disappears.

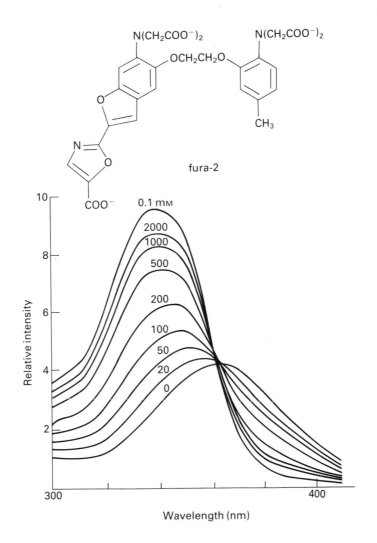

Fig. 9.35. The excitation spectrum of fura-2 shifts upon binding calcium ions. Here is shown a typical set of excitation scans obtained under ionic conditions appropriate for mammalian cytoplasm as fura-2 is titrated through a series of increasing $[Ca^{2+}]$ concentrations. Free $[Ca^{2+}]$ values, indicated in nanomolar quantities except for the '0.1 mM' curve, were well buffered with EGTA. The excitation maxima shifts toward 340 nm as $[Ca^{2+}]$ increases. The emission was monitored at 505 nm throughout. Note the isosbestic point at 360 nm. Inset shows the structure of fura-2. (From Tsien & Poenie, 1986.)

The non-injected sister blastomere continues to divide and its calcium concentration remains low. The calcium dye, fura-2, was injected into the single cell before division to ensure an equal distribution. This experiment shows the critical role of calcium in mitotic spindle dynamics.

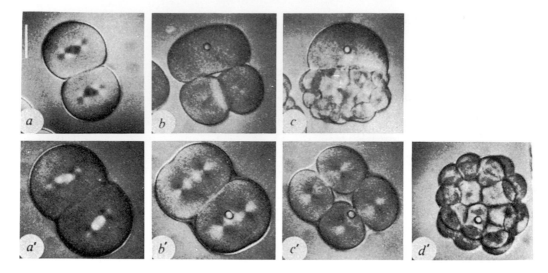

Fig. 9.36. Effect of the microinjection of antibody to 46K protein into a two-cell sea urchin embryo at mitosis. (a) The birefringence of the two sets of mitotic aparatus in the blastomeres immediately before the injection indicates the mitotic stage. (b) Several minutes after the injection (indicated by the oil drop) the birefringence disappears and the injected cell is arrested, whereas the untreated sister-blastomere has divided. (c) The injected blastomere remains undivided for the time of observation, but the control blastomere has developed into a blastomere-like stage. (a'–d') The result of the injection of unrelated monoclonal antibodies. No effects on the development are found. (Scale bar, 20 μm.)

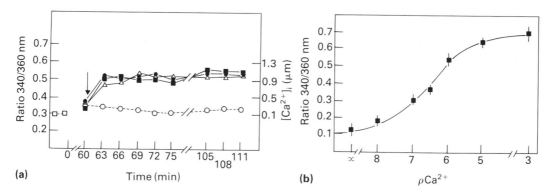

(a)

(b)

Fig. 9.37. Increase of $[Ca^{2+}]_i$ in a sea urchin blastomere after injection of antibody to 46K protein. (a) Time course of the calcium increase. The injected cell was centered in the measuring field of the photometer. The data for the free Ca^{2+} on the right ordinate are derived from the calibration curve shown in b. Ca^{2+} is measured in the unfertilized egg (□), at metaphase shortly before microinjection (arrow), and after microinjection at the times indicated; the results of three different experiments are shown. $[Ca^{2+}]_i$ can be seen to rise slightly at mitosis, but increases sharply within two minutes after injection of the antibody solution. The time course of the $[Ca^{2+}]_i$ fluctuations in a non-injected control cell is represented by ○. (b) Relationship between the ratio of fura-2 fluorescence (340/360) nm) and Ca^{2+}. *In vivo* calibration: sea urchin eggs were microinjected with fura-2 to give final concentrations of 10–20 μM. The surrounding sea water was replaced by a solution similar to the intracellular composition: 220 mM potassium glutamate, 500 mM glycine, 10 mM NaCl, 2.5 mM MgCl₂, pH 6.9, containing varying amounts of Ca^{2+}. The pCa^{2+} of this solution was controlled by varying the ratio of Ca-EGTA/EGTA. 1–2 μM ionomycin were added to the injected eggs to allow equilibration of intra- and extracellular Ca^{2+} (data from Hafner & Petzelt, 1987).

9.13   Absorption spectrophotometry

The instrument

Newton's fundamental discovery (Chapter 8) that white light on refraction is dispersed into its constituent colours is the basis of the prism spectrophotometer. The parallel beam of white light from a source will be dispersed into its constituent colours by the prism and these will be focused at different points on the focal plane of the lens (Fig. 9.38). The chosen wavelength can be made to fall on the slit by rotating the prism and, after passing through the sample, the intensity of the light is measured by means of a photocell. Spectrophotometers are required to work over a wide wavelength range and there are basically four technical problems to overcome.

(a)   *Suitable light source*

In order to overcome the first difficulty, two lamps have to be used to cover the necessary wavelength range. Usually they are a tungsten lamp which produces a broad spectrum to cover the range 1000–320 nm, and a hydrogen lamp, which produces narrow emission lines, is used from 320 to 190 nm.

(b)   *Means of producing a pure spectrum*

Either prisms or diffraction gratings (Section 8.14) are used to obtain a pure spectrum. Light of the required wavelength is obtained by rotating the prism, or grating, and the rotating head carries a calibrated wavelength scale.

(c)   *Material with which to make lenses and absorption cells*

As glass absorbs strongly in the lower wavelength (ultraviolet) regions, the prism, lenses and test cells have to be made from silica.

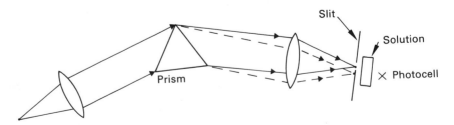

Fig. 9.38. Spectrophotometer, with prism monochromator. Blue light (dashed) is refracted more than red (solid) and so, with the prism mounted on a rotating table, a narrow range of wavelengths can be selected by the slit.

(d)   *Means of detecting light transmitted through specimen solution*

Photocells are normally used to detect the light passing through the sample and the output from these is amplified and the result displayed on a scale or chart paper calibrated in terms of percentage transmission or optical density.

Theoretical treatment

Absorption is a matter of probability and in order that the statistical laws hold we must assume this treatment will be confined to dilute solutions.

Suppose we have a light beam of intensity I which, after passing through the test cell of width ΔX, holding a solution, falls on a photocell which measures the transmitted energy $(I - \Delta I)$. If we carried out this experiment we would find that the amount of light absorbed, ΔI was proportional to I and this is reasonable because the more photons there are in the incident beam, the greater the probability of one being absorbed. Provided the width ΔX was small, ΔI would also be proportional to ΔX. ΔI would also depend on the substance in the solution, e.g. on whether the material had an absorption band at the wavelength used. The constant for the material μ is called the *absorption coefficient*.

Hence we can write

$$\Delta I = - \mu I \Delta X.$$

Now if we allow ΔX to become infinitely thin

$$dI = -\mu I\, dx$$

or

$$dI/dx = -\mu I$$

and integrating this equation from $x = 0$ to $x = x$ gives

$$I_x = I_0 \exp(-\mu x) \tag{9.6}$$

where I_0 is the intensity incident on the cell, and I_x is the intensity at a distance x from the surface.

As absorption is a probability phenomenon then the more absorbers there are in the light path the more will be absorbed. For dilute solutions the absorption coefficient will be proportional to the concentration, C, i.e.

$$\mu = \beta C$$

where β is known as the *extinction coefficient*. Sometimes the *transmittance* is measured and this is simply the ratio I_x/I_0. Because of the very

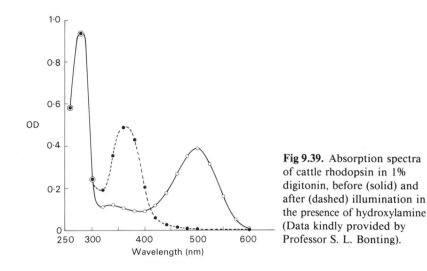

Fig 9.39. Absorption spectra of cattle rhodopsin in 1% digitonin, before (solid) and after (dashed) illumination in the presence of hydroxylamine (Data kindly provided by Professor S. L. Bonting).

wide range of transmittances possible, e.g. from 0.1 to 100%, it is usually more convenient to display the result on a logarithmic scale and in this way the *optical density*, *D*, is measured. *D* is defined by the equation

$$D = \log_{10} \frac{I_0}{I_x}$$

and as
$$\mu = \frac{1}{x} \log_e \frac{I_0}{I_x}$$
(9.7)

then
$$D = \frac{\mu x}{2.303}$$
(9.8)

and
$$C = 2.303 \, D/\beta x.$$
(9.9)

In absorption spectrophotometry a further quantity is often used, the *molar extinction coefficient* ϵ_{mol}, defined by the relationship

$$\epsilon_{mol} = \frac{D}{Cx}.$$
(9.10)

The units invariably used are not SI but because their use is so widespread they will be given here: *C* has units mol litre^{-1} and the cell path length *x* is in centimetres, ϵ_{mol} then has the units litre mol^{-1} cm^{-1}.

Problem 9.2

Photoreceptor cells (rod outer segments, Fig. 13.11) were isolated from cattle retinae, solubilized in digitonin and the absorbance (optical density) read before and after bleaching (Fig. 9.39). Given that the total dry weight concentration of the receptors was 2.7 mg ml^{-1}, that the molar extinction coefficient of rhodopsin at 500 nm is 41 000 l mol^{-1} cm^{-1}, that the molecular weight of rhodopsin is 40 000 and that a 1 cm spectrophotometer cell was used, show that rhodopsin represents a significant fraction of the total outer segment dry weight.

9.14 Molecular motion in membranes

Polarization optics combined with absorption spectrophotometry can be used to study molecular motion as well as alignment. This has been demonstrated most elegantly with rod photoreceptor cells which

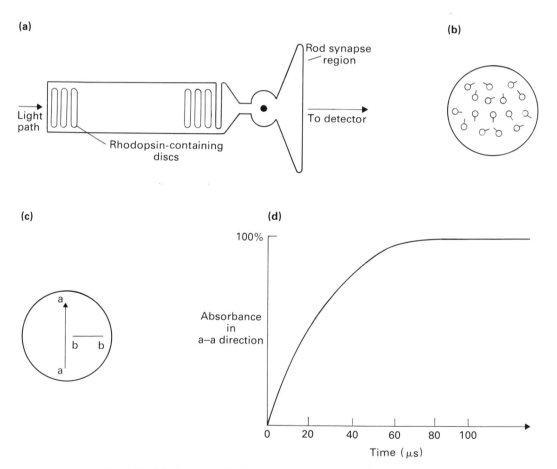

Fig. 9.40. (a) Diagram of light path through a single photoreceptor cell. (b) End-on view of a single disc. The rhodopsin molecules are randomly oriented. (c) Direction of plane of polarization of bleaching flash (a–a). Rhodopsin molecules with their chomophore lined up in this direction have a high probability of being bleached. Those with chromophores lined up in the b–b direction have zero probability of bleaching. (d) Absorbance of 500 nm light plane-polarized in the a–a direction, following a brief, intense bleaching flash in the a–a direction. Note that the half-time for recovery is approximately 20 μs.

contain the light-sensitive molecule rhodopsin within their disc membrane (see Fig. 13.11). Brown and Cone (1972) mounted a retina on a holder in a microspectrophotometer so that a narrow beam of light was passing along the long axis of a single photoreceptor cell (Fig. 9.40a). They measured the absorption of 500 nm, plane-polarized light and found that the absorption coefficient was independent of the plane of polarization, thus showing that the rhodopsin chromophores were randomly aligned in the plane of the disc. However, they found that they could induce dichroism by bleaching the discs with plane polarized light (Fig. 9.40c).

All the chromophores lined up in the a–a direction absorb light, while those in the b–b direction do not. Hence, if the 500 nm absorbance is quickly measured after a very brief flash of light, with a polarizing filter in front of the detector, it is found to be very low with the filter in the a–a direction, but high in the b–b direction. However, the absorbance in the a–a direction rapidly builds up with time as the unbleached rhodopsin chromophores rotate in the plane of the disc. After about 100 μs, the absorbances in the a–a and b–b directions are equal (Fig. 9.40).

The rhodopsin chromophore (retinaldehyde or vitamin A) lies buried deep within the hydrophobic regions of the rhodopsin molecule (Fig. 9.41) represented here, for simplicity, as a prolate spheroid. It actually consists of a single polypeptide chain which crosses the membrane five times. The chromophore lies along the plane of the disc membrane and spins rapidly in the dark. Interestingly, polarization optics can also be used to show that the chromophore becomes disengaged from the polypeptide chain on illumination and rotates through 90° to lie perpendicular to the plane of the disc.

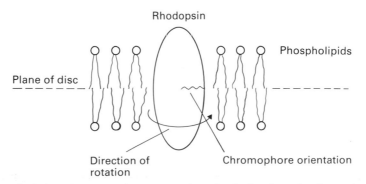

Fig. 9.41. Diagrammatic representation of a rhodopsin molecule rotating in the plane of the disc. The half-time for the motion is rapid ($\approx$ 20 μs) showing that the molecule rotates freely. The rhodopsin molecule is also free to move laterally in the plane of the disc. This motion may play an important role in the transduction process where rhodopsin has to interact with several other types of molecules between the capture of the incoming light photon and the consequent electrical activity of the receptor.

References

Beeston, B.E.P., Horne, R. W. & Markham, R. (1972) *Electron Diffraction and Optical Diffraction Techniques.* North-Holland, Amsterdam.

Epstein, H.T. (1963) *Elementary Biophysics: Selected Topics.* Addison-Wesley, Reading, Mass.

Hafner, M. & Petzelt, C. (1987) Inhibition of mitosis by an antibody to the mitotic calcium transport system. *Nature*, **330**, 264–6.

Harding, C.V., Chylack, L.T., Susan, S. R., Lo, W-K. & Bobrowski, W.F. (1982) Elemental and ultrastructural analysis of specific human lens opacities. *Investigative Ophthalmology and Visual Science*, **23**, 1–13.

Maurice, F., Meny, L. & Tixier, R. (1980) *Microanalysis and Scanning Electron Microscope.* Editions de Physique, Orsay, France.

Thomas, J.B. (1965) *Primary Photoprocesses in Biology.* North-Holland, Amsterdam.

Tsien, R.Y. & Poenie, M. (1986) Fluorescence ratio imaging: a new window into intracellular ionic signalling. *Trends in Biochemical Science*, **11**, 450–6.

Further reading

Agar, A.W., Alderson, C. & Chescoe, D. (1974) *Principles and Practice of Electron Microscope Operation.* North-Holland, Amsterdam.

Bradbury, S. (1976) *The Optical Microscope in Biology.* Arnold, London.

Brown, P.K. (1972) Rhodopsin rotates in the visual receptor membrane. *Nature*, **236**, 35–8.

Campbell, I.D. & Dwek, R.A. (1984) *Biological Spectroscopy.* Benjamin Cummings Inc., Menlo Park, CA, USA.

Chandler, J.A. (1977) *X-ray Microanalysis in the Electron Microscope.* North-Holland, Amsterdam.

Clayton, R.K. (1965) *Molecular Physics in Photosynthesis.* Blaisdell, New York.

Cone, R.A. (1972) Rotational diffusion of rhodopsin in the visual receptor membrane. *Nature*, **236**, 39–43.

De Robertis, E.D.P. & De Robertis, E.M.F. (1980) *Cell and Molecular Biology.* Holt-Saunders, Philadelphia.

Holz, H.M. (1977) Worthwhile facts about fluorescence microscopy. *Zeiss Information* K 41–005–E (Oberkochen, W. Germany).

Horne, R.W. (1979) The formation of virus crystalline and paracrystalline arrays for electron microscopy and image analysis. *Advances in Virus Research*, **24**, 1–50.

Misell, D.L. (1978) *Image Analysis, Enhancement and Interpretation.* North-Holland, Amsterdam.

Morgan, A.J. (1985) *X-ray Microanalysis in Electron Microscopy for Biologists.* Oxford University Press, Oxford.

Reimer, L. (1985) *Scanning Electron Microscopy — Physics of Image Formation and Microanalysis.* Springer, Berlin.

Whittingham, C.P. (1974) *The Mechanism of Photosynthesis.* Arnold, London.

Wilson, H.R. (1966) *Diffraction of X-rays by Proteins, Nucleic Acids and Viruses.* Arnold, London.

10 Electricity

10.1 Introduction

At present, the two most important areas of interest for a biologist studying electricity are *electrostatics* and *current electricity*. The former is required to understand the interaction of charged species, ions and molecules, at the molecular and cellular level. The latter is essential in order not only to understand the electrical signalling processess that take place along both animal and plant membranes, but also to use intelligently the vast range of electronic equipment available to biologists of all interests. The practical basis of the design and use is dealt with in Chapter 11.

10.2 Basic concepts

The study of electricity is concerned with the behaviour of charged particles, electric fields and magnetic fields. It is unfortunate that there is no simple intuitive model for the easier understanding of these terms. An *electric field* is defined in terms of its action on a charge and an *electric charge* is defined in terms of its behaviour in an electric field. The tautological nature of these concepts is extended to the definition of a magnetic field, which can neither be produced nor detected without the aid of moving charges. Charge has a positive or negative sign. Two bodies, A and B, with charges of the same sign exert a repulsive force on one another (Fig. 10.1) and consequently work must be done to bring them together. The converse holds with bodies of opposite sign.

10.3 Coulomb's law

The experimental relationship found for charged spheres (Coulomb's Law) is

$$F = \frac{q_1 q_2}{4 \pi K \epsilon_0 x^2} \tag{10.1}$$

where F is the magnitude of the force between the spheres and x is the distance between the centres of the charged bodies. The units of charge are coulombs (C) and charge is an integral multiple of the charge on a single electron ($e = 1.6 \times 10^{-19}$ C). When q_1 and q_2 have the same sign

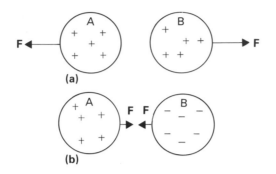

(a)

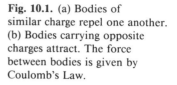

(b)

Fig. 10.1. (a) Bodies of similar charge repel one another. (b) Bodies carrying opposite charges attract. The force between bodies is given by Coulomb's Law.

the force exerted is a repulsive one (Fig. 10.1). ϵ_0 is a coefficient called the *absolute permittivity* of free space and has the value 8.85×10^{-12} $C^2 \ m^{-2} \ N^{-1}$. K is the *relative dielectric constant* of the medium separating the charges and has the value of 1 for a vacuum and 80 for water, for example (Section 10.22).

Like gravitational attraction this force is of the 'action at a distance' type, making itself felt without the presence of a material connection between A and B. It is not known why this is possible, it is simply an experimental fact. It is useful to think of each of the charged bodies as modifying the space around it, and setting up an electric field.

10.4 Electric field

In Fig. 10.1 (a), if body B is removed, then A is still said to exert an *electric field* at the point P (Fig. 10.2). To show that an electric field exists at P, one simply places a small test charge (q) there and, if a force is exerted, a field is said to exist at P.

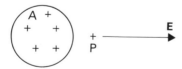

Fig. 10.2. The fact that a force is exerted on a small test charge placed at P indicates the presence of an electric field there. The field vector (**E**) is in the same direction as the force vector when the test charge is positive.

The field vector **E** is defined by the relation

$$\mathbf{E} = \frac{\mathbf{F}}{q} \tag{10.2}$$

when $F = 1$ N and $q = 1$ C, then $E = 1 \ N \ C^{-1}$.

Suppose a small charge q moves a distance s from a to b under the influence of a field **E**, then the work done W in moving this distance is given by

$$W = \int_b^a \mathbf{F}.\mathrm{d}s = q \int_b^a \mathbf{E}.\mathrm{d}s \tag{10.3}$$

Note that the work done is the integral of a vector product, i.e.

$$W = q \int_b^a \mathbf{E} \cos \theta \, \mathrm{d}s \qquad \text{(Appendix 2)}.$$

It can be shown that the work done in an electric field is conserved, i.e. $\int_b^a \mathbf{E}.\mathrm{d}s$ is independent of the path taken. When a force has this property (for example, gravitational force), the work done on the particle is the difference in potential energy of the particle between the end point and the starting point, i.e.

$$W = U_b - U_a,$$

i.e. $\quad U_b - U_a = q \int_b^a \mathbf{E}.\mathrm{d}s \qquad (10.4)$

where U_a and U_b are the initial and final potential energies respectively.

10.5 Electric potential

Instead of dealing directly with the potential energy of a charged particle it is useful to introduce the more general concept of energy per unit charge. This quantity is called *potential, V*. Thus for a particle of charge q and potential energy U_p,

$$V = \frac{U_p}{q} \qquad (10.5)$$

and thus from equation (10.4).

$$V_b - V_a = \int_b^a \mathbf{E}.\mathrm{d}s. \qquad (10.6)$$

$V_b - V_a$ is the potential difference between the two points a and b. It has the units of $\mathrm{J\,C^{-1}}$. One $\mathrm{J\,C^{-1}}$ has the special name of one *volt*. A value can be assigned to the potential at a single point only when some arbitrary reference point is selected at which the potential is zero. In the case of the potential difference across a cell membrane, for example, it is usually convenient to regard the outside medium as the reference point.

10.6 Summary

The interaction of two systems of charges can be characterized in several ways (Fig. 10.3).

1   There is a force vector $\mathbf{F}$ on a charge q in the vicinity of a system of charges S.
2   There is a vector field $\mathbf{E}$ at a due to the presence of the system S.

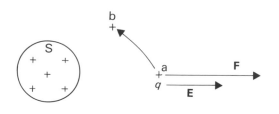

Fig. 10.3. The interaction of two systems of charges can be described in several ways. (See text for explanation.)

3 Work is done when q moves from a to b and the work is independent of the path taken.

4 There is an electrical potential difference between the points a and b, given by the work done in moving a unit positive charge from a to b.

10.7   The cathode ray oscilloscope

The cathode ray oscilloscope is widely used in electrophysiology when rapidly changing electrical signals are studied, e.g. nerve action potentials. The mode of operation of the cathode ray tube itself (Fig. 10.4) provides a simple example of the movement of charges, electrons, under the influence of electric fields.

The whole interior of the tube is highly evacuated. The cathode at the left is raised to a high temperature by the heater, and electrons, cathode rays, evaporate from its surface. The *accelerating anode* is maintained at a high positive potential, V, relative to the cathode so that there is an electric field directed from right to left between the

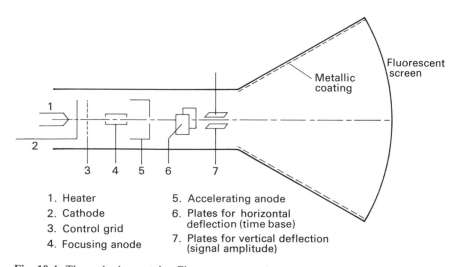

1. Heater
2. Cathode
3. Control grid
4. Focusing anode
5. Accelerating anode
6. Plates for horizontal deflection (time base)
7. Plates for vertical deflection (signal amplitude)

Fig. 10.4. The cathode ray tube. Electrons are accelerated towards the screen by an electric field between the cathode (2) and anode (5).

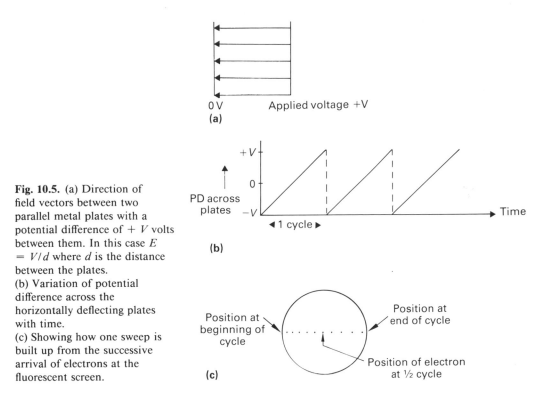

Fig. 10.5. (a) Direction of field vectors between two parallel metal plates with a potential difference of $+V$ volts between them. In this case $E = V/d$ where d is the distance between the plates.
(b) Variation of potential difference across the horizontally deflecting plates with time.
(c) Showing how one sweep is built up from the successive arrival of electrons at the fluorescent screen.

anode and cathode. This field is confined to the cathode–anode region and electrons passing through the hole in the anode travel with a constant velocity from the anode to the fluorescent screen. The function of the control grid is to regulate the number of electrons that reach the anode and hence the brightness of the spot on the screen.

The complete cathode–anode assembly is called the *electron gun*. The electrons then pass between two pairs of deflecting plates, the first of which controls the horizontal deflection of the beam (Fig. 10.5a). The sweep speed is controlled by the potential difference across the plates which varies in time with a sawtooth waveform (Fig. 10.5b). The electrons arrive initially at the left side of the screen and, as the field decreases, they are deviated less and so they arrive at the centre of the screen when the field across the plates is zero halfway through the cycle. The field then increases in the opposite sense. At the end of the cycle the sequence will be repeated if the oscilloscope trigger system (Fig. 10.6) is set to the internal trigger mode. The horizontal sweep can also be started by an external trigger pulse or it can be initiated by a change in the level or slope of the voltage across the vertical plates. The field across the vertical plates is controlled by the potential difference across the source under investigation and hence a two-dimensional image is formed on the screen. This can be photographed for detailed study later. Fig. 10.6

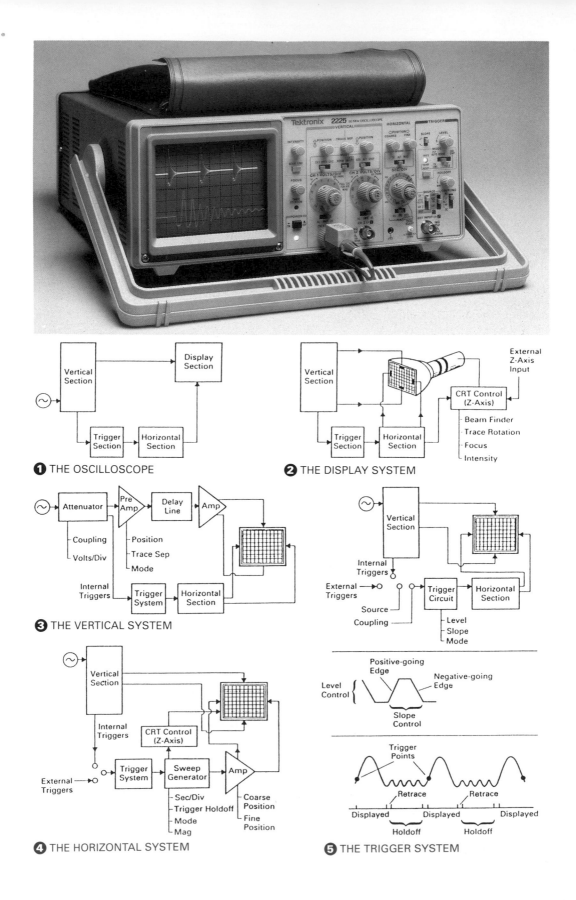

1 THE OSCILLOSCOPE

2 THE DISPLAY SYSTEM

Beam Finder
Trace Rotation
Focus
Intensity

3 THE VERTICAL SYSTEM

Coupling
Volts/Div

Position
Trace Sep
Mode

4 THE HORIZONTAL SYSTEM

Sec/Div
Trigger Holdoff
Mode
Mag

Coarse Position
Fine Position

5 THE TRIGGER SYSTEM

Level
Slope
Mode

Positive-going Edge
Negative-going Edge
Slope Control

Trigger Points
Retrace   Retrace
Displayed   Displayed   Displayed
Holdoff   Holdoff

Fig. 10.6. (a) Photograph of front view of a typical oscilloscope. This version permits two input voltages (CH1 and CH2) to be placed across the vertical deflection plates (plates 7 in Fig. 10.4) and they can be switched rapidly 'on' and 'off' during one cycle. Two traces can therefore be displayed on the screen during a cycle and the switching (chopping) is sufficiently rapid so that both traces appear continuous. The voltages 1 and 2 are normally measured with respect to earth, but they can also be subtracted (differential input), when only a single trace appears on the screen.
More detailed descriptions of these controls and others are given in the manufacturer's handbook (Tektronix Inc). Block diagrams 1–5 illustrate the operation of a typical oscilloscope. The input signal is connected to the vertical plates and diagram 1 shows that, if desired, this voltage can be used to trigger the oscilloscope and control the display. The horizontal plates (6 in Fig. 10.4) are controlled by an amplifier producing a 'sawtooth' waveform (Fig. 10.5) and this amplifier is in turn switched by the trigger system. Diagram 2 also shows the controls for the electron beam (z-axis). Diagrams 3 and 4 show the interrelations between the control circuits for the vertical and horizontal displays, while diagram 5 shows that this particular oscilloscope permits a wide variety of trigger modes so that certain parts of the input signal can be studied in greater detail. (Photographs and diagrams kindly provided by Tektronix Inc.)

---

shows a typical oscilloscope with block diagrams of the various control circuits.

10.8 Electric fields and sense organs

Certain types of electric fish, of which *Gymnarchus* is a striking example, are able to locate objects by sensing changes in the electric fields set up by the fish themselves (Fig. 10.7 *overleaf*).

The fish have stacks of modified tissue called electric organs set up in a regular array along their bodies and through these maintain a considerable electrical potential difference between the head and tail. The field is pulsed by the fish and 25 ms is the average length of a pulse of which there are about 7 or 8 per second. The skin of the fish probably acts as an insulator except in certain regions near the head where jelly-filled pits provide conducting paths for the electric field. These pits have some type of sense organ at their base and it is these receptors that can detect very small changes in the field distribution at the head. With its detectors a fish can detect the presence of a conducting rod in its tank even when it is only 2 mm in diameter. These fish are able to live in muddy streams where the use of eyes for catching prey or avoiding predators is impossible.

Until recently it was believed that only certain fish and amphibia possessed electroreceptors. However, it has now been established that the higher vertebrate, the Australian diving platypus, can both locate and avoid objects on the basis of electric fields that are only of the order of 50 mV m^{-1}. The electroreceptive organ is located in the animal's characteristic bill (Scheich *et al.*, 1986).

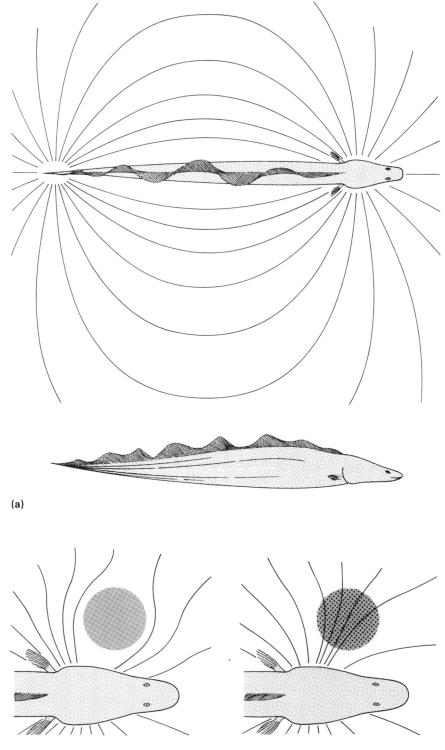

(a)

(b)

Fig. 10.7. (a) Showing the electric field of *Gymnarchus* and the location of electric generating organs. Each electric discharge from organs in the rear portion of the body makes the tail negative with respect to the head. Most of the electric sensory pores or organs are in the head region. The undisturbed electric field resembles a dipole field, as shown, but is more complex. The fish responds to changes in the distribution of electric potential over the surface of its body. The conductivity of objects affects the distribution of potential. (b) Objects in the electric field of *Gymnarchus* distort the lines of current flow. The lines diverge from a poor conductor (left) and converge toward a good conductor (right). Sensory pores in the head region detect the effect and inform the fish about the object (from Lissman, 1963)

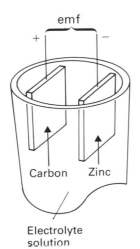

Fig. 10.8. Primary cell.

10.9 Electromotive force

A body with a considerable excess of electrons has a strong negative charge and one with a deficit has a positive charge. Between them there exists a field so long as they are separated. If some material allowing a free movement of electrons is placed between two bodies carrying different charges, electrons move through this conductive path. If the movement of electrons is to be maintained, a continuous surplus must be generated to replace electrons moving away through the conductor. The surplus can be maintained chemically, as in a battery, or electromagnetically as in a generator. These systems are termed sources of *electromotive force* (emf).

One common example of a battery is the *primary cell* (Fig. 10.8) which consists of zinc and carbon plates immersed in an electrolyte solution of ammonium chloride. Zinc goes into solution as the divalent ion Zn^{2+}, leaving two electrons behind so that a negative charge is left on the zinc plate. The solution becomes positively charged and so does the carbon plate in contact. When a wire connects the two terminals, *current* flows and this is maintained by a continual dissolving of the zinc plate. The emf developed in this type of cell is about 1.5 V. When drawing circuit diagrams, a source of emf is represented by | **ı**, the long line denoting the positive (carbon) terminal and the shorter line the negative (zinc) terminal.

If batteries are connected in *parallel* (Fig. 10.9a) there is more area for chemical action and a more powerful current is available but the potential available remains the same. However, when they are connected in *series* (10.9b), the voltage steps are additive and a higher overall voltage is available.

10.10 Electric current

Electric current, as mentioned above, is the flow of charge and is defined by the relationship

$$I = \frac{\mathrm{d}q}{\mathrm{d}t} \tag{10.7}$$

where the current I has the units $\mathrm{C\,s^{-1}}$ which are given the special name *ampères*. The current is the rate at which coulombs are passing a given point in a system. As charge has a sign, current will have a sign. It is conventional to regard the direction of flow of current as the direction of flow of positive charge, whatever the sign of the charge of the moving particles. In the case of current flow across a cell membrane, a further arbitrary definition of sign must be made. The convention that will be used here is that a positive membrane current is a positive current flowing into the cell.

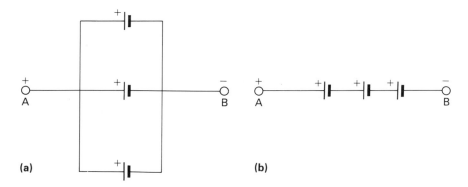

Fig. 10.9. (a) Batteries connected in parallel, each of emf 1.5 V the total open circuit emf ($V_A - V_B$) is still 1.5 V, but more current will be available. (b) When the batteries are connected in series, the open circuit emf is now 4.5 V, but the total current available will be the same as for one battery alone.

10.11 Electrical conductance

Certain substances allow currents to pass relatively easily and these are termed *conductors*. Other substances offer much more resistance to the flow of current and these are termed *insulators*. A part of a circuit whose resistance is negligible is called a *short circuit*; one whose resistance is effectively infinite is called an *open circuit*.

In the case of a metallic solid, the outer, valency, electrons are loosely held to the atomic nuclei and can move freely about the crystal lattice as a sort of electron gas. Such a substance is a good conductor. If such a substance is heated up, electrons can be driven from the surface. Heated cathodes are used to provide free electrons in valves, cathode ray tubes, X-ray tubes and electron microscopes.

Another type of conduction is *ionic conduction*. Here the mobile charged particles are ions which are much larger particles than electrons and which have either a negative, *anion*, or positive, *cation*, charge. These can only move significantly if the medium is of relatively low viscosity. This is the mechanism of conduction in electrolyte solutions and probably in biological membranes.

Insulators are substances with few mobile electrons or ions. However, there is no sharp distinction from conductors. Crystalline solids showing relatively poor electronic conduction are termed *semiconductors*. Examples are germanium, silicon and selenium. These have a number of important practical applications, e.g. thermistors, photoelectric cells, rectifiers and transistors (Chapter 11).

Insulators are often exposed to enormous electric fields and the materials of which they are composed are subject to enormous stress. As the field is increased there comes a point when there is a physical breakdown of the insulator, *dielectric breakdown,* and a surge of current flows through the insulator. The maximum field an insulator can support is termed its *dielectric strength*. The potential difference across a cell membrane is usually about 0.1 V, inside negative and, as membranes are only of the order of 5 nm thick, the field across the membrane is thus about 2×10^7 V m^{-1}. Artificial membranes undergo dielectric breakdown at a potential difference of about 0.25V, suggesting that the cell membrane is operating at a point relatively near breakdown.

10.12 Ohm's Law

In Fig. 10.10(a), AB represents a conductor and the experimental relationship between the current through the conductor and the potential drop across it was discovered by Georg Ohm.

$$I = \frac{1}{R}(V_A - V_B) \qquad (10.8)$$

or

$$I = G(V_A - V_B) \qquad (10.9)$$

where R is the resistance across AB and has units $V\ A^{-1}$, also given the special name *ohm* (shorthand symbol: Ω) G is the conductance and has units Ω^{-1}. The SI unit of conductance is the *siemen*. If the current–voltage curve of a conductor is linear (Fig. 10.10b), the behaviour is described as *ohmic*. The curves obtained from many biological membranes, e.g. from nerve membranes, are non-ohmic and are termed *rectifying*. In this case the resistance to current passing in one direction is greater than the resistance to current flowing in the opposite direction. The conductance, or resistance, is in fact a function of potential.

10.13 Temperature and resistance

The resistance of an ordinary metallic conductor increases with increasing temperature. Thus a resistor placed in a resistance measuring circuit can, after calibration, be used to measure temperature. Semiconductors show a particularly large temperature effect with the resistance actually decreasing with increasing temperature. Such a semiconductor device is known as a *thermistor* and because of its small size, and thus rapid response time, the thermistor has important biological applications.

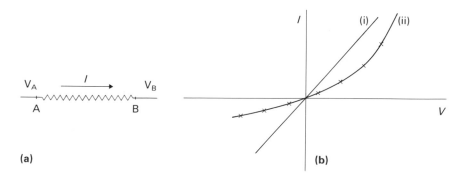

Fig. 10.10. (a) AB represents a resistor. V_A and V_B are the potentials at A and B respectively when a current I flows. (b) Current–voltage curves for (i) an ohmic resistor, e.g. a copper wire and (ii) a non-ohmic resistor, e.g. squid nerve membrane.

227 ELECTRICITY

10.14 Heating effects of currents: Joule's Law

Consider a conductor that has current I flowing through it (Fig. 10.10a). In the time interval dt, a quantity of charge dq, given by

$$dq = Idt,$$

enters the portion of the circuit at terminal A and in the same time an equal quantity of charge leaves through B. As there is a transfer of charge dq from potential V_A to V_B, work dW must be done, given by

$$dW = dq(V_A - V_B) \tag{10.10}$$

$$= Idt \, V_{AB}. \tag{10.11}$$

The power input, or rate of energy dissipation, is given by

$$P = dW/dt$$

$$= IV_{AB} = I^2R \tag{10.12}$$

where R is the resistance between A and B.
Equation (10.12) is called *Joule's Law* and the power input is mainly dissipated in the form of heat.

What are the units of power in this case?

$$P = I \times V = C\,s^{-1} \times J\,C^{-1}$$

$$= J\,s^{-1} = watts.$$

10.15 Circuit equations

Joule's Law can be used to derive certain equations that are useful in solving problems on electrical circuits.

In Fig. 10.11 points a and b are at the same potential as they are connected together by a non-resistive wire. However, a potential difference exists across the resistance R. Suppose the current flowing round the circuit is I, then heat is developed in the external resistor at a rate I^2R. It is also found that every source of emf has an internal resistance, usually low, called r. Hence the rate of heat development in the cell is I^2r.

The total power supplied by the emf is EI and this must be equal to the heat energy dissipated in the resistors

$$\therefore EI = RI^2 + rI^2 \tag{10.13}$$

or

$$I = \frac{E}{R + r}.$$

Generally

$$I = \frac{\Sigma E}{\Sigma R}.$$ (10.14)

Note that a convention must be adopted for the direction of current flow and the usual one is that current is flow of positive charge and so travels from the positive terminal of a source of emf, round the external circuit, to the negative terminal.

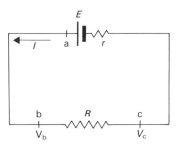

Fig. 10.11. An emf E of internal resistance r is driving current through the external resistor R.

10.16　Potential difference between points in a circuit

The rate at which circulating charge gives up energy to the portion of the circuit between a and b (Fig. 10.12) is IV_{ab}. This is the power input to this portion of the circuit supplied by sources of emf in the remainder of the circuit (not shown). Power is also supplied *by* the first source of emf (E) and power is supplied *to* the second source (E'). Heat is developed in the various resistors.

Fig. 10.12. See text for explanation.

Equating power input and output

$$IV_{ab} + EI = E'I + (R + r + r')I^2.$$ (10.15)

Equation (10.15) can be stated in a more general form

$$V_{ab} = \sum RI - \sum E.$$ (10.16)

Careful attention must be paid to algebraic signs. Direction from a to b is always considered positive. Currents and emfs are positive if their direction is from a to b. Resistances are always positive.

If a and b coincide, $V_{ab} = 0$

$$\therefore I = \frac{\Sigma E}{\Sigma R}.$$ (10.14)

10.17   Series and parallel connection of resistors

Series (Fig. 10.13a)

When the resistors are connected in series, the current through each is the same; hence

$$V_{ab} = IR_1 + IR_2 + IR_3 .$$

From Fig. 10.13(b),

$$V_{ab} = IR$$

where R is the equivalent resistance. Hence

$$R = R_1 + R_2 + R_3. \tag{10.17}$$

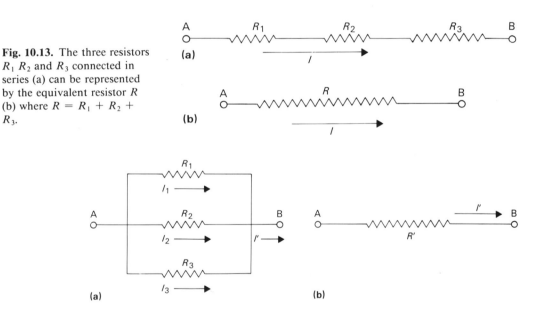

Fig. 10.13. The three resistors R_1 R_2 and R_3 connected in series (a) can be represented by the equivalent resistor R (b) where $R = R_1 + R_2 + R_3$.

Fig. 10.14. The three resistors in parallel (a) can be represented by the equivalent resistor R' (b) where $1/R' = 1/R_1 + 1/R_2 + 1/R_3$.

Parallel

When the resistors are connected in parallel (Fig. 10.14a) I' is split up into three paths, i.e.

$$I' = I_1 + I_2 + I_3$$

but the potential difference across the resistors is the same V_{ab} as they have common connections. Now

$$I_1 = \frac{V_{ab}}{R_1}, \quad I_2 = \frac{V_{ab}}{R_2} \text{ and } I_3 = \frac{V_{ab}}{R_3}$$

but from Fig. 10.14(b)

$$I = \frac{V_{ab}}{R'}$$

$$\therefore \quad \frac{V}{R'} = \frac{V}{R_1} + \frac{V}{R_2} + \frac{V}{R_3}$$

so $\quad \dfrac{1}{R'} = \dfrac{1}{R_1} + \dfrac{1}{R_2} + \dfrac{1}{R_3}.$ $\qquad\qquad$ (10.18)

10.18 Kirchoff's rules

Not all networks can be reduced to simple series or parallel arrangements (Fig. 10.15) and in order to analyse such networks another approach, devised by Gustav Kirchoff, has to be used. He introduced the terms *branch point* and *loop* and they are defined as follows:
(i)   A *branch point* occurs when three or more conducting paths meet, e.g. points a and b in Fig. 10.15
(ii)   A *loop* is any closed conducting path.
    The basic rules for the solution of network problems are:
(i)   *Point rule*. The algebraic sum of the currents *towards* any branch point is zero, i.e.

$$\sum I = 0. \qquad\qquad (10.19)$$

Note that if there are n branch points in a network, there are only $n - 1$ independent point equations.
(ii)   *Loop rule*. The algebraic sum of the emfs in any loop equals the algebraic sum of the IR products in the same loop, i.e.

$$\sum E = \sum IR. \qquad\qquad (10.20)$$

    Some sign convention must also be adopted in conjunction with rules and a useful one is:
(i)   *Point rule convention*. Current is considered positive when its direction is towards a branch point.
(ii)   *Loop rule convention*. The clockwise direction is considered positive and all currents and emfs in this direction are positive.

Problem 10.1

Write down the equations to solve the network in Fig. 10.15.

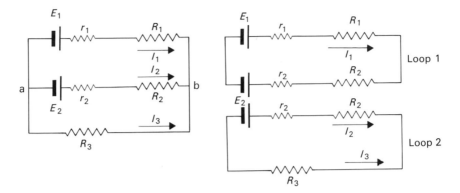

Fig. 10.15. Example of a network that cannot easily be analysed in terms of series and parallel arrangements of resistors. a and b are the two branch points and loops 1 and 2 are the two loops.

Answer

There are two branch points, a and b and so there is only one independent point equation.

$$I_1 + I_2 + I_3 = 0.$$

considering loop 1

$$E_1 - E_2 = I_1 r_1 + I_1 R_1 - I_2 r_2 - I_2 R_2$$

and loop 2

$$E_2 = I_2 r_2 + I_2 R_2 - I_3 R_3 .$$

There are now three independent equations to solve for the three unknown currents.

Problem 10.2

A *Wheatstone bridge* (Fig. 10.16) is an instrument for determining the value of an unknown resistor. It consists of four resistors, a source of emf, and a current detector (*ammeter*) arranged as shown. The known variable resistor R_2 is adjusted until there is a zero reading on the ammeter and the bridge is then said to be in balance. The reader should easily be able to show that the magnitude of the unknown resistor R_1 is given by

$$R_1 = R_2 (R_4 / R_3)$$

where R_3 and R_4 are known resistors of fixed value.

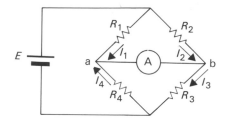

Fig. 10.16. The Wheatstone bridge. R_3 and R_4 are fixed, known resistors; R_2 is a known variable; R_1 is the unknown resistor. When the bridge is in balance there is no current flowing through the ammeter A.

10.19   Membranes — potentials and conductances

Membrane potentials are set up by the diffusion of ions down their *electrochemical gradients*. Although the actual mechanism of the passage of ions through membranes is poorly understood, we can obtain some understanding of the problem by the application of elementary thermodynamics.

Suppose we have a membrane separating two KC1 solutions of different concentrations $(K_i > K_0)$ and suppose the membrane is permeable only to potassium ions (Fig. 10.17). The potassium ions will diffuse down their concentration gradient, taking positive charge with them so that the outside phase will acquire a positive potential with respect to the inside. This potential will tend to impede the subsequent movement of cations and a point will be reached when there will no longer be a net diffusion of potassium ions from inside to out. At this point the driving force on the ions due to the concentration gradient will be exactly balanced by the driving force due to the potential difference.

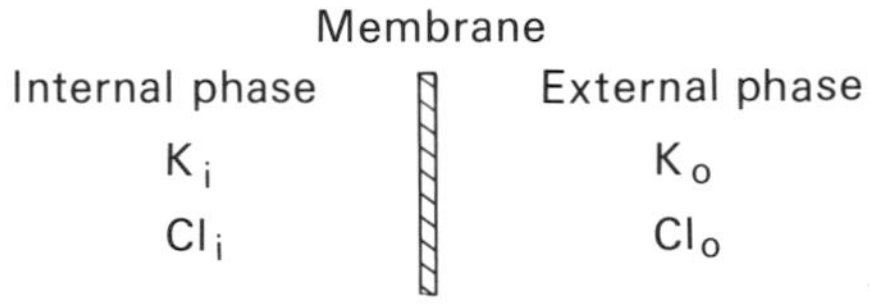

Fig. 10.17. A membrane separating two phases. It is assumed that only potassium ions can pass through.

	Membrane	
Internal phase		External phase
K_i		K_o
Cl_i		Cl_o

We can obtain the magnitude of this potential, called the *equilibrium* or *Nernst potential*, by considering the work that has to be done to move a small quantity of K^+ ions from the outside phase to the inside. From elementary thermodynamics the work required to move δn moles through the membrane against the concentration gradient is given by

$$\delta W = \delta n\, R\, T \log_e \frac{[K_i]}{[K_0]} \tag{10.21}$$

where R is the gas constant, T is the absolute temperature, and K_i and

K_0 the concentrations of potassium in the internal and external phases respectively.

Now the work done to move δn moles against a potential difference E is

$$\delta W = \delta nZFE \qquad (10.22)$$

where Z is the valency of the ion (equiv. mol^{-1}), F is the Faraday (96 500 C per equiv.), and E is the potential difference across the membrane. The reference for potential is taken as the external solution.

At equilibrium, there will be no net work done, and so

$$\delta nZFE = -\delta nRT \log_e \frac{[K_i]}{[K_0]}$$

and

$$E = -\frac{RT}{ZF} \log_e \frac{[K_i]}{[K_0]} \qquad (10.23)$$

$$= -25 \log_e \frac{[K_i]}{[K_0]} \times 10^{-3} \text{ V at } 20°C$$

$$= -58 \log_{10} \frac{[K_i]}{[K_0]} \times 10^{-3} \text{ V}$$

The actual membrane potential difference in a biological system will contain contributions from all of the ionic species present, and the contribution from each will depend on the *permeability* of the species. The more permeable the ion, the more it will contribute to the potential. In most animal cells, for example, chloride and potassium are much more permeable than sodium, so they contribute most to the potential.

In analysing complex electrical currents in the membrane it is often helpful to have a model to work from and membrane biophysicists make great use of the equivalent circuit (Fig. 10.18). The diffusion potentials are represented by batteries and the actual contribution which these make to the overall membrane potential will be determined

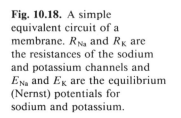

Fig. 10.18. A simple equivalent circuit of a membrane. R_{Na} and R_K are the resistances of the sodium and potassium channels and E_{Na} and E_K are the equilibrium (Nernst) potentials for sodium and potassium.

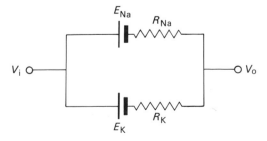

by the ease with which ions can pass through their appropriate channel in the membrane. The lower the permeability of an ion, the greater will be the resistance of that channel. It is left as an exercise to show that when there is no net current across the membrane, the potential across it is given by:

$$V_i - V_0 = \frac{E_{Na} R_K}{R_{Na} + R_K} + \frac{E_K R_{Na}}{R_{Na} + R_K}. \tag{10.24}$$

The resistance of the cell membrane is normally measured by passing a current pulse ΔI across the membrane via two electrodes, one internal and one external, and then measuring the resultant change in potential ΔV across another pair. If the area of cell membrane is A, the current through a unit area is $\Delta I / A$ and the resistance is the change in potential divided by the current, i.e.

$$R_m = A \frac{\Delta V}{\Delta I}. \tag{10.25}$$

Resistance in this case has the units of Ωm^2 and the resistance of most animal cell membranes is in the region of $10^{-1} \Omega m^2$. The conductance is therefore $10 \ \Omega^{-1} \ m^{-2}$ or $10 \ Sm^{-2}$.

Problem 10.3

The resistance of most cell membranes is of the order of $10^3 \ \Omega \ cm^2$ $(10^{-1} \ \Omega \ m^2)$. We wish to measure the transmembrane potential in an amphibian oocyte of diameter $10^{-3} \ m$ and in order to do this a

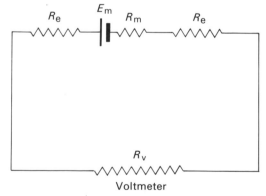

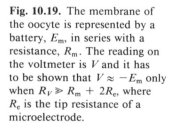

Fig. 10.19. The membrane of the oocyte is represented by a battery, E_m, in series with a resistance, R_m. The reading on the voltmeter is V and it has to be shown that $V \approx -E_m$ only when $R_V \gg R_m + 2R_e$, where R_e is the tip resistance of a microelectrode.

microelectrode with a tip resistance of $10^6 \ \Omega$ is inserted. The tip resistance of the reference microelectrode is also $10^6 \ \Omega$. Show that a *volt-*

meter with a high internal resistance, R_V, has to be used for an accurate measurement of the potential.

The circuit for the system is given in Fig. 10.19.

10.20   Capacitance

The discussion so far has dealt with steady electrical currents. When the voltage is not constant, as is the case when it is first applied, or if it is an alternating voltage, e.g. in power lines, then some additional phenomena are seen.

For any current to flow in a circuit there must be a complete path of finite resistance through which the electrons can flow and return to their starting place. However, when a voltage is first applied to a conductor there will be a movement of electrons into the conductor even in the case of an open circuit. What happens is that the excess of electrons in the conductor produces an electric charge which builds up with still more electrons until it balances the applied voltage. Then with no voltage gradient there will be no further movement of electrons, and hence no current. The situation described could be called a one-plate capacitor (Fig. 10.20).

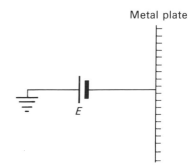

Fig. 10.20. Current only flows for a short time when the battery is connected to the metal plate because an opposing voltage is set up equal in magnitude but opposite in sign to the emf.

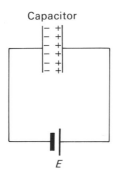

Fig. 10.21. See text for explanation.

If two sheets of metal are placed close together (Fig. 10.21), but separated from contact by a good insulator, they form a *parallel plate capacitor* and a movement of electrons will occur when a voltage is applied across the two plates. However, on one side of the circuit electrons will flow out of the plate leaving a net positive charge while, on the other, electrons will flow in and the whole applied potential will be concentrated in a rather steep gradient across the insulator between the plates. This insulator, called the *dielectric*, can by its dimensions and dielectric properties affect the ultimate density of electrons on the plates required to equalize the applied voltage. Thus the charge on the capacitor, which means the actual number of electrons in excess or deficit on the plates, varies not only with the applied voltage, but also with the area of the plates and the type of dielectric between the plates. It can be shown that the capacitance of a parallel plate condenser is given by

$$C = \frac{K \varepsilon_0 A}{d} \tag{10.26}$$

where K is the relative dielectric constant, ε_0 is the absolute permittivity, A is the area of the plates, and d the distance between them.

The term *capacitance* denotes a capacity to store charge and the relationship between stored charge and applied potential is given by

$$Q = CV. \tag{10.27}$$

When Q is in coulombs and V is in volts, the units of C are *farads*. A capacitor of one farad will store enough electrons for a one volt applied potential to require a net charge of one coulomb to counteract the applied potential. This is a very large number of electrons, and capacitors in the microfarad and picofared range are more commonly found in biology. The capacitance of most cell membranes is of the order of $10 \ \mathrm{mF \ m^{-2}}$.

In a steady direct current situation, once the capacitor is charged to the maximum for the applied potential, no more current will flow and the circuit resembles an open circuit. When the current is reversing its direction rapidly, as in alternating current circuits, it is obvious that the electrons will be able to move back and forth from one plate of the condenser to the other, with each change in polarity of the applied voltage, so there will be a current flow in the circuit even though there is an insulating dielectric between the plates. If the alternations are slow, there will be time for an equilibration to be reached at each point in the cycle and the number of electrons on the plates will create a counter-potential which will stop the flow of current. However, if the voltage reverses very rapidly and if the electrical resistance is high enough to limit the current flow, there will never be enough electrons on the plate of the condenser to counteract the applied potential effectively. In such a case the current is limited by the resistance in the circuit alone. Thus the limiting effect of the condenser on the current depends both on its capacitance C and also on the frequency f of alternations of the voltage. The limiting effect is called the *capacitive reactance* X_c, and is measured in ohms, just as resistance is:

$$X_c = \frac{1}{2\pi f C}. \tag{10.28}$$

A capacitor is usually thought of in terms of metal plates separated by an insulating medium, or dielectric. The concept is equally applicable to the double-leaflet structure of cell membranes where $C \approx 10$ mF m^{-2}, or 1 μF cm^{-2} using the units most likely to be met.

10.21 Energy stored in a capacitor

As charging a capacitor involves the movement of charge from the plate at a lower potential to the plate at a higher potential, work must be done.

Suppose that a total of q units of charge have been added to one plate and that the potential difference between the plates is V, the work done dW in transferring further charge dq will be given by

$$dW = V\,dq \tag{10.29}$$

and as $\quad q = C V$

$$dW = \frac{q}{C}\,dq.$$

The total work W in increasing the charge from 0 to Q is

$$W = \int_0^Q \frac{1}{C}\,q\,dq \tag{10.30}$$

$$= \frac{1}{2}\frac{Q^2}{C} = \frac{1}{2}CV^2. \tag{10.31}$$

This work done is stored as energy in the capacitor and is released when it discharges.

10.22 Dielectric constant

The dielectric constant is an important property of matter and the value for many materials can readily be determined by sandwiching them between the plates of a capacitor. The capacitance of this system is then measured by a current pulse technique (Section 10.23) and the dielectric constant can be obtained (from equation 10.26) if the geometry of the capacitor (A/d) is known.

K, known as the relative dielectric constant, is nearly 1 for most gases, but for some substances which are strong dipoles the value is much greater, e.g. for water $K = 80.4$ (20°C). The physical basis of the high dielectric constant of dipolar media is that when the potential difference is applied the dipoles rotate, aligning themselves with the electric field (Fig. 10.22). The orientated dipoles act to neutralize the charge on the plates, thus increasing the capacity of the system. The bulk of the energy stored in the capacitor is the energy required to orientate the dipoles from the random state. If the field is allowed to decay, the dipoles randomize and, providing there is no hysteresis, the energy is recovered. This randomization or relaxation of the dipoles is effectively a flow of current across the system.

In the case of the cell membrane the capacitance per unit area (10 m F m^{-2}) defines the ratio $K\varepsilon_0/d$. If $d \approx 5$ nm, then $K \approx 5$. A value as low as this might be regarded as surprising for a system mainly composed of lecithin, which is a strong dipole. However, there is evidence that the lecithin bilayer has a considerable degree of order, although it would be incorrect to regard it as in a frozen, orientated state. The term *liquid crystal* is used for such a state as the cell membrane. The degree of order restricting the free rotation of the dipoles would be responsible for a low dielectric constant in such a bilayer.

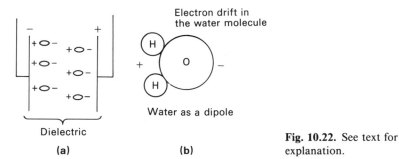

Electron drift in the water molecule

Water as a dipole

Dielectric

(a)

(b)

Fig. **10.22.** See text for explanation.

10.23 Circuits containing capacitance and resistance (Fig. 10.23)

First let us derive a very useful relationship between current, capacitance and potential difference, starting with equation (10.27):

$$Q = CV$$

and, on differentiating,

$$dQ/dt = Cd V/dt.$$

as capacitance is independent of time. Hence

$$I = Cd V/dt. \tag{10.32}$$

When the switch s is closed a current *I* will flow to charge up the condenser (Fig. 10.23a)

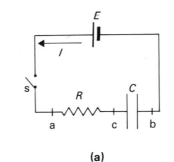

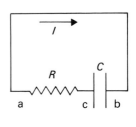

Fig. 10.23. (a) When the switch s is closed, a current I will flow round the circuit to charge up the capacitor.
(b) After charging up, if the ends ab are short-circuited, current will flow and the capacitor will be discharged.

(a)

(b)

$$E = V_{ac} + V_{cb}$$

$$\therefore \quad V_{cb} = E - IR$$

$$\text{but} \quad I = C\,\mathrm{d}V_{cb}/\mathrm{d}t$$

$$\therefore \quad V = E - RC\,\mathrm{d}V/\mathrm{d}t.$$

V_{cb} will simply be referred to as V

$$RC\,\mathrm{d}V/\mathrm{d}t = E - V$$

$$\frac{\mathrm{d}V}{E - V} = \frac{1}{RC}\,\mathrm{d}t$$

and, on integration

$$-\ln\,(E - V) = \frac{1}{RC}t + M$$

where M is the integration constant. Or

$$(E - V) = \exp\left(-\frac{t}{RC}\right)\exp\,(-M).$$

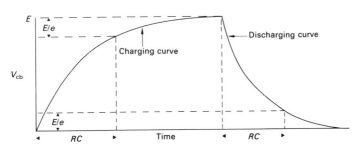

Fig. 10.24. Charging curve for the potential across the capacitor.

Now when $t = 0$, $V = 0$

$$\therefore \quad \exp\,(-M) = E$$

$$\text{i.e.} \quad E - V = E\,\exp\,(-t/RC)$$

$$\therefore \quad V = E\,(1 - \exp - t/RC). \tag{10.33}$$

RC (Fig. 10.24) is referred to as the *time constant* of the circuit and has units of seconds.

Now suppose that after charging up the condenser to the potential E we let it discharge through the resistance R, i.e. disconnect the source of emf and connect the points a and b together (Fig. 10.23b)

$$V_{ac} + V_{cb} = 0$$

i.e. $- IR + V_{cb} = 0$

where current is flowing from c to a across the resistor, and

$$I = -C \frac{dV_{cb}}{dt}$$

as the capacitor is discharging. Therefore

$$V = -RC \frac{dV}{dt}$$

writing V for V_{cb} again.

If we integrate once more and take the boundary condition $V = E$ when $t = 0$, the solution to this equation is

$$V = E \exp(-t/RC). \tag{10.34}$$

In this case the time constant RC is the time for the potential to decrease to $1/e$ (0.37) of its original value (Fig. 10.24). The capacitance of a cell membrane, for example, can be determined by passing a square pulse of current (Fig. 10.25a) through the membrane by means of a pair of glass microelectrodes. The potential difference across the membrane is measured simultaneously by means of a further electrode pair and the voltage response shows typical charging and discharging characteristics.

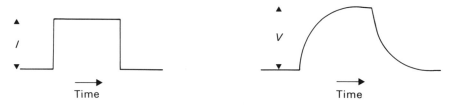

Fig. 10.25. (a) Square current pulse applied. (b) Voltage response observed.

As the membrane resistance is given by the ratio V_{max}/I, the membrane capacitance (mF m^{-2}) can be calculated from the RC characteristic of the membrane.

10.24   Capacitors in series and parallel

Capacitors in series (Fig. 10.26)

If the left plate of C_1 receives a charge $+Q$, then a charge $-Q$ is induced on the right plate and $+Q$ appears on the left plate of C_2 etc.

$$Q = C_1 V_{ab} = C_2 V_{bc} = C_3 V_{cd}$$

and

$$V_{ad} = V_{ab} + V_{bc} + V_{cd}$$

and if C represents the equivalent capacitance, i.e. the capacitance of the single capacitor that would become charged with the same charge Q when the potential difference across its terminals is V_{ad}.
Then

$$V_{ad} = \frac{Q}{C}$$

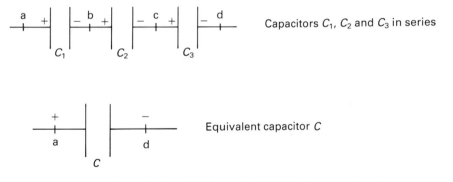

Capacitors C_1, C_2 and C_3 in series

Equivalent capacitor C

Fig. 10.26. (a) Capacitors C_1, C_2 and C_3 in series. (b) Equivalent capacitor C.

$$\therefore \frac{Q}{C} = \frac{Q}{C_1} + \frac{Q}{C_2} + \frac{Q}{C_3}$$

$$\therefore \frac{1}{C} = \frac{1}{C_1} + \frac{1}{C_2} + \frac{1}{C_3}. \tag{10.35}$$

Capacitors in parallel (Fig. 10.27)

The potential across each capacitance is V_{ab}, but the charge induced on each is different.

$$Q_1 = C_1 V, \quad Q_2 = C_2 V, \quad Q_3 = C_3 V.$$

The total charge on the parallel network is

$$Q = Q_1 + Q_2 + Q_3.$$

Defining the equivalent capacitance as the one for which

$$Q = CV_{ab}$$

$$CV_{ab} = C_1 V_{ab} + C_2 V_{ab} + C_3 V_{ab}$$

$$\therefore \ C = C_1 + C_2 + C_3. \tag{10.36}$$

Compare the equivalent capacitor equations 10.35 and 10.36 with the equivalent resistors for the series and parallel arrangements (equations 10.17 and 10.18).

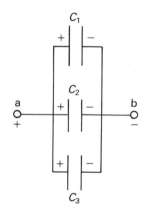

Fig. 10.27. See text for explanation.

Problem 10.4

A membrane can be represented by the equivalent circuit shown in Fig. 10.28. E_{Na} and E_K are the sodium and potassium Nernst potentials and R_{Na} and R_K represent the resistances of the sodium and potassium channels in the membrane. C is the membrane capacitance and E_m, the transmembrane potential, is given by $V_i - V_o$.

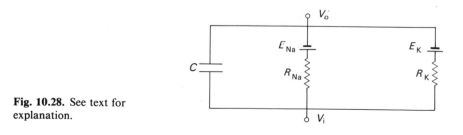

Fig. 10.28. See text for explanation.

(a) The following data have been obtained for squid nerve in the resting state. $Na_0 = 440$ mM, $K_0 = 20$ mM, $Na_i = 50$ mM, $K_i = 400$ mm; $E_m = -60$ mV (inside negative) and the total membrane resistance equals 1 kΩ cm². Show that $E_{Na} = +55$ mV, $E_K = -75.5$ mV and calculate the ratio of the resistances of the sodium and potassium channels.

(b) At the height of the action potential, $E_m = +40$ mV. Calculate the total resistance at the height of the action potential, assuming that only R_{Na} has had time to change, and compare your computation with the experimental value of 40 Ω cm². Why is the computed resistance the larger?

(c) The membrane capacitance is 1 μF cm^{-2}. How many moles of ions move through the membrane in order to discharge it during an action potential? Take $ZF = 96\,500$ C mol^{-1}. See also Duncan & Croghan (1973) for an equivalent circuit analysis of photoreceptor potentials.

Problem 10.5

(a) 'Solve' the following circuits by calculating the value of a single resistor which would give the same current flow in the circuit.

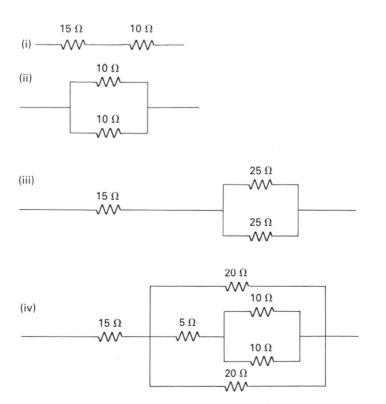

(b)   Find an equivalent value of capacitance which could be used to substitute for the following arrangements:

(i)

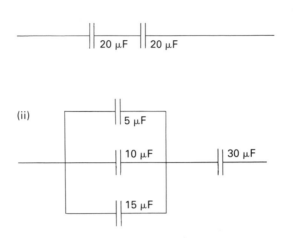

(ii)

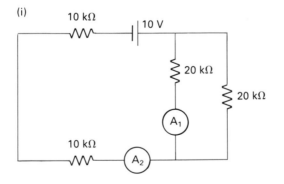

(c)   Calculate the current at each point indicated by an ammeter (A)
Note: You may assume that the ammeters have negligible internal resistance.

(i)

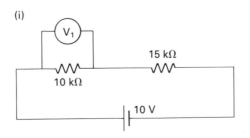

(d)   Calculate the voltage that would register on each voltmeter.

(i)

(ii)

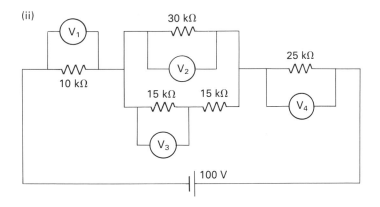

You may assume that the internal resistance of each voltmeter is very high, $>10^9$ Ω.

Problem 10.6

An α-particle is a nucleus of doubly ionized helium. It has a mass m of 6.68×10^{-27} kg and a charge q of $+2e$ or 3.2×10^{-19} C. Compare the force of electrostatic repulsion (in air) between two α-particles with the force of gravitational attraction between them.

Problem 10.7

Torpedo occidentalis is a large electric fish believed to use its electricity in attack and defence rather than for obstacle location or social purpose. A typical individual would be capable of producing potentials up to 220 V (measured at the electric organ) driving pulses of current of 15 A through its sea-water environment. Pulses are typically 2×10^{-3} s in duration and occur in bursts during which the firing frequency is 200 Hz. Assuming the pulses to be of square waveform with voltage and current in phase, calculate: (a) the rate, in kilowatts, at which electrical work is done by the electric organs during a pulse; (b) the rate at which electrical work is done by the electric organs, averaged throughout a burst of pulses; (c) the total work done in one pulse; (d) the cost of this energy if bought from the Electricity Board at 5.5 pence per kilowatt-hour.

References

Duncan, G. & Croghan, P.C. (1973) Excitation and adaptation in the cephalopod retina: an equivalent circuit model. In *Biochemistry and Physiology of Visual Pigments* (ed. H. Langer). Springer, Berlin.

Lissman, H.W. (1963) Electric location by fishes. In *From Cell to Organism*, Freeman, San Francisco.

Scheich, H., Langner, G., Tidemann, C., Coles, R.C. & Guppy, A. (1986) Electroreception and electrolocation in platypus. *Nature,* **319**, 401–2.

Further reading

Aidley, D.K. (1989) *The Physiology of Excitable Cells* (3rd edn). Cambridge University Press.

Hodgkin, A.L. (1964) *The Conduction of the Nervous Impulse.* Liverpool University Press.

Jarman, M. (1970) *Examples in Quantitative Zoology.* Arnold, London.

Katz, B. (1966) *Nerve, Muscle and Synapse.* McGraw-Hill, New York.

Sears, F.W. & Zemansky, M.W. (1964) *University Physics.* Addison-Wesley, Reading, Mass.

Strong, P. (1973) *Biophysical Measurements.* Tektronix, Beaverton, Oregon.

11 Electronics: Amplifiers

11.1 Introduction

Electrical processes are involved at practically all stages of any biological system, however complex or simple, since alterations in the movement of ions across membranes allow us to see, hear, write, move, remember etc. We also require electronic instruments to understand the molecular mechanisms involved in the *control* of all of these processes. For example, in the processes of vision, only a few quanta of light are required to give a visual sensation. The quanta are absorbed by the photopigment rhodopsin contained within the photoreceptor cells. An internal transmitter substance (cGMP) is released which alters the movement of sodium ions through the photoreceptor cell plasma membrane. There is therefore a *transduction* of light energy into electrical energy and *amplification* occurs as the very small energy content of a photon alters the movement of many millions of ions through the membrane (Fig. 11.1). The blockade of the movement of sodium ions results in a change in the photoreceptor membrane potential and this is electrically processed in the retina before it is relayed to the visual cortex. Our visual perception is largely made up of a very great number of 'bits' of information stored in different cell layers in the visual cortex. Command signals (feedback) can also be sent from the visual cortex to the eye to modify the visual input (Fig. 11.2). For example, if the incoming illumination is very high, then signals are sent via the optic nerve to the iris muscles to reduce pupil size.

In order to measure the very small electrical signals produced at all stages of the information processing, very finely-tipped electrodes have to be placed in or near the cells responsible. Since such electrodes have a large resistance, the electronic devices used to amplify the signals must

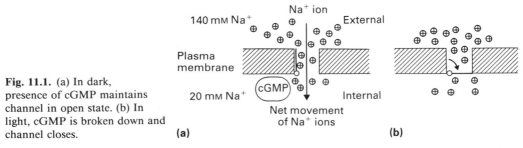

Fig. 11.1. (a) In dark, presence of cGMP maintains channel in open state. (b) In light, cGMP is broken down and channel closes.

247

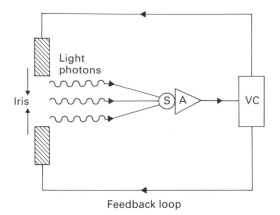

Iris

Light photons

Feedback loop

Fig. 11.2. Example of feedback control system in vision. The currents produced by the photons are summed in the photoreceptor/amplifier system and the output is sent via the retina to the visual cortex (VC). The pupil diameter is controlled by the feedback loop so that when the ambient light intensity is high the iris moves to reduce the diameter.

have a high *input impedance* (Chapter 10). In order to understand how such amplifiers work, we have to understand the electronic processes of conduction in solids.

11.2   Conduction in solids

(a)   Conductors

Substances that are good heat *conductors* are also good conductors of electricity and in both situations the transfer of energy is effected by the movement of electrons through the solid matrix. The specific resistance of a good conductor is of the order of 10^{-8} Ωm.

(b)   Insulators

In substances that are good *insulators* the electrons are tightly bound to the constituent atoms and so there are few available to carry either electrical current or heat energy through the material. Resistances of insulators are of the order of 10^4 Ωm.

(c)   Semiconductors

Materials with intermediate resistance values (10^{-1} Ωm) are termed *semiconductors*. The resistance of these materials has interesting properties as it is sensitive to changes in ambient temperature and, more importantly, it is extremely sensitive to the presence of certain impurities within the matrix of the semiconductor. In order to understand how these useful properties arise and are utilized in the manufacture of electronic components such as *diodes* and *transistors,* it is necessary to understand a little of the structure of solids and the mechanism of conduction.

Fig. 11.3. (a) Simplified energy level diagram showing the overlap that occurs when two or more atoms interact. (b) Arrangement of energy levels in an atom which is part of a larger ensemble of similar atoms (e.g. in a semiconductor). The energy difference (forbidden zone) is very small in conductors and very high in insulators. In semiconductors it has an intermediate value.

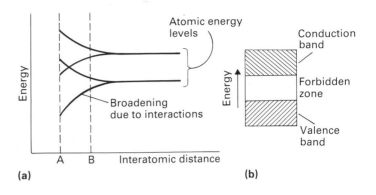

11.3 Electronic energy levels in solids

In a single atom the allowed electronic energy levels are widely separated but in the solid state, such as in a crystal, the atoms are in close proximity and the electrons are strongly influenced by neighbouring nuclei. The permissible energy levels are therefore broadened into bands (Fig. 11.3).

In the situation at A (Fig. 11.3a) there is a complete overlap of energy levels, while at B there are energy levels separated by a band of energy which the electrons cannot occupy, called the *forbidden zone*. The lowest available energy band is termed the *valence band* and the next available energy band is termed the *conduction band*.

In an insulator all the electrons are in the valency band and the conduction band is empty. The two bands are separated by a very wide forbidden zone. In an energy versus interatomic distance graph for an insulator the energy levels never overlap, whereas for a good conductor they do. In semiconductors the valence band is narrow and is of the order of the kinetic energy possessed by an electron at room temperature (kT). In a semiconductor the proportion of electrons in the valence and conduction bands depends on temperature. At 0 K all of the energy levels in the valence band are occupied and the material is an insulator.

Semiconductor materials

Silicon and germanium are the most widely used semiconductor elements and both are tetravalent. In a pure crystal of silicon, for example, each atom is covalently bonded to four neighbouring atoms by the sharing of valence electrons (Fig. 11.4). At 0 K, all of the valence electrons are firmly bound to the nucleus of their particular atom, while at room temperature the thermal energy of one of the valence electrons may become greater than the binding energy of the nucleus. The

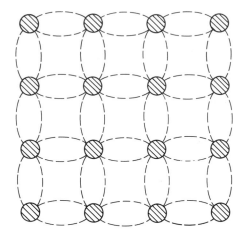

 Si atom

Valence bond
formed by atoms
sharing two electrons

Fig. 11.4. Diagram of crystal
lattice structure of pure silicon.

(a)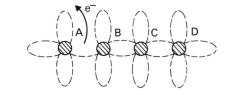

(b)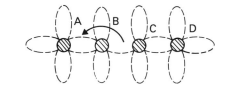

(c)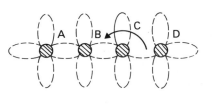

Direction of 'hole' migration

Fig. 11.5. A valence electron
shared between A and B gains
sufficient thermal energy to
join the lattice conduction band,
creating a positively charged
'hole'. An electron shared
between B and C can move to
fill this hole, creating another
unpaired valence bond.
Similarly, an electron shared
between C and D can migrate
to the left.

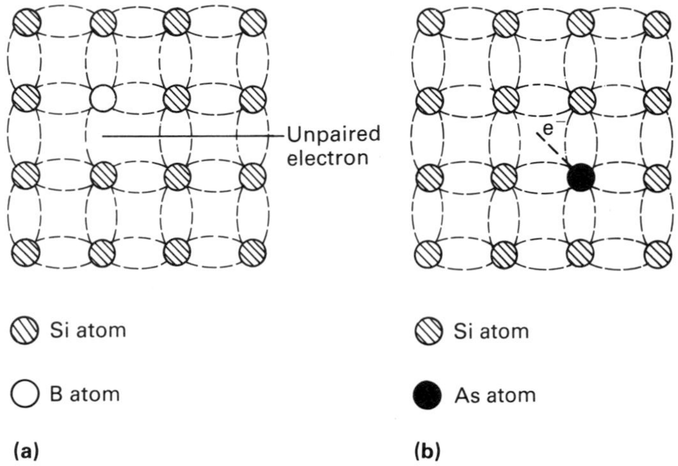

Fig. 11.6. (a) Effect of introducing a trivalent impurity into the silicon lattice. A 'hole' is created between the boron atom and one of its silicon neighbours. This 'hole' can migrate through the crystal as before (Fig. 11.5). Since positive holes are formed, this type of semiconductor is termed a p-type. (b) Effect of introducing a pentavalent atom (e.g. arsenic) into the silicon lattice. The arsenic atom has five electrons in its outer shell and only four are required to bond-share with neighbouring silicon atoms. The excess electron is free to migrate in the crystal lattice and hence this type of doping forms a negative, or n-type, semiconductor.

—Unpaired electron

⊘ Si atom

○ B atom

(a)

⊘ Si atom

● As atom

(b)

electron then leaves the atom and diffuses freely through the crystal. The atom is left with an unpaired electron and it carries a net positive charge (Fig. 11.5). This creates an attractive electronic force on the surrounding electrons and one of them may leave their atom to fill the vacancy. This hole-filling process can occur many times in the crystal and, while an electron migrates from atom D to atom A, the 'hole' migrates in the opposite directon.

In a pure or *intrinsic* semiconductor the number of conduction electrons and holes are equal. It is possible, however, to increase either the number of holes or electrons in a tetravalent semiconductor by *doping* it with small amounts of trivalent (Fig. 11.6a) or pentavalent (Fig. 11.6b) impurities respectively.

11.4 Junction diode

Small pieces of p and n type semiconductors can be joined to form a *diode* (Fig. 11.7). The charged species are free to diffuse across the junction and a 'barrier' potential difference (E_B) of ≈ 10 mV is set up. As a result of this diffusion, a 'depletion layer' is formed in the region of the junction since holes and electrons have diffused from the p and n regions respectively. The width of this layer is of the order of microns.

In the forward bias mode, the source of emf provides electrical energy which tends to move the electrons and holes across the junction in the same direction as the concentration gradient. In the reverse bias

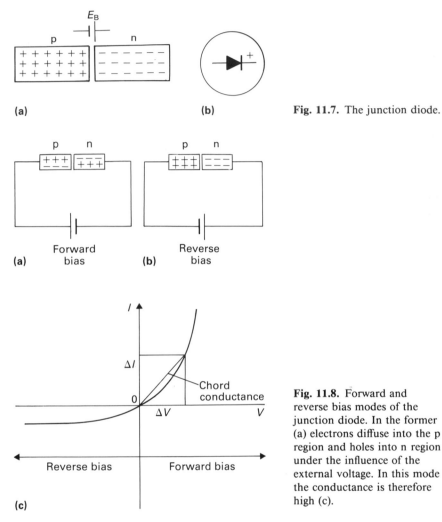

Fig. 11.7. The junction diode.

Forward bias (a)   Reverse bias (b)

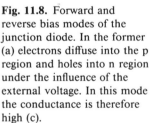

Fig. 11.8. Forward and reverse bias modes of the junction diode. In the former (a) electrons diffuse into the p region and holes into n region under the influence of the external voltage. In this mode the conductance is therefore high (c).

direction, only the intrinsic carriers are free to carry current in the direction dictated by the external source and so the conductance is much lower (Fig. 11.8c).

The conductance, calculated either by the slope (dI/dV) or chord ($\Delta I/\Delta V$) method, increases with forward bias voltage and this non-linear, or non-ohmic behaviour is shared by many types of biological membranes.

11.5 Zener diode

If a very high voltage is set up across the narrow p–n junction then electrons are stripped from their atoms and, because of their very high kinetic energy, can remove other electrons from other atoms. This

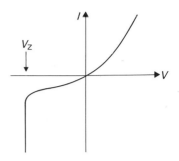

Fig. 11.9. Current–voltage characteristics of the Zener diode.

greatly increases the number of electron–hole pairs and the conductance of the diode increases dramatically (Fig. 11.9). The voltage at which this occurs is termed the Zener voltage (V_Z) and is characteristic of the materials making up the p–n junction.

Voltage regulation

If an accurately defined voltage is required for a circuit, then a Zener diode can be used to regulate a varying voltage supply since, in the breakdown region, the voltage across the diode is constant for relatively large current fluctuations.

11.6 Transistors

A junction transistor consists of a sandwich of semiconductor materials, either in the form n–p–n or p–n–p. The inner region is called the

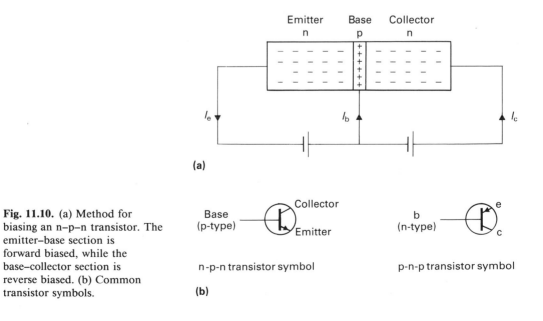

Fig. 11.10. (a) Method for biasing an n–p–n transistor. The emitter–base section is forward biased, while the base–collector section is reverse biased. (b) Common transistor symbols.

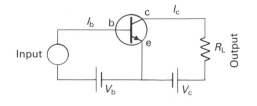

Fig. 11.11. Common-emitter configuration for n–p–n transistor. The input signal is placed between the base and emitter, while the output is recorded across the resistor R_L.

base (b) while the outer segments are the *collector* (c) and *emitter* (e) respectively.

The symbols for both types of transistors are shown in Fig. 11.10 and they are differentiated by the direction of current flow at the emitter. For example, in the n–p–n arrangement the emitter–base section is forward biased (Fig. 11.10a) and there will be a large flow of electrons from emitter to base (I_e). These electrons rapidly reach the collector region by diffusion and, as the base–collector section is reverse biased, they are accelerated into the collector region to form the collector current (I_c). Some of the electrons combine with holes in the base region and so I_c is smaller then I_e. The difference has to be made up by the external base current (I_b).

$$I_e = I_b + I_c.$$

Small changes in base current (ΔI_b) produce very large changes in collector current (ΔI_c) and so the current gain in the system ($\Delta I_c/\Delta I_b$) can be very large.

The usual configuration for a transistor amplifier is the common-emitter configuration (Fig. 11.11). The input is connected between the transistor base and the biasing voltage (V_b) while the voltage output is obtained across the load resistor (R_L) connected between the collector and the biasing voltage (V_c).

The relationships between base current and voltage and collector current and voltage define the transistor characteristics (Fig. 11.12).

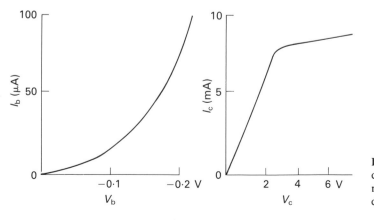

Fig. 11.12. Current–voltage characteristics for a typical n–p–n transistor in the common-emitter configuration.

The input resistance, $\Delta V_b/\Delta I_b$ is of the order of 200Ω, but the output resistance ($\Delta V_c/\Delta I_c$) can be extremely high, of the order of several $M\Omega$, and so R_L can have high values without influencing the transistor characteristies. The gain or amplification in this configuration is given by the ratio: output voltage/input voltage and again it can be very large.

11.7   Field effect transistors (FET and MOSFET)

The FET is again formed by a semiconductor sandwich (e.g. p–n–p), but in this case the n-type material forms a narrow channel between two thin wafers of p-type material (Fig. 11.13).

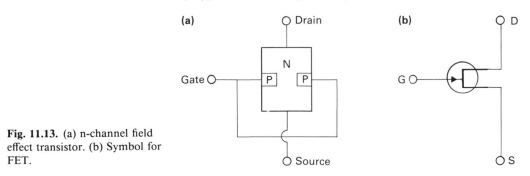

Fig. 11.13. (a) n-channel field effect transistor. (b) Symbol for FET.

Electrons can normally pass freely through the channel from source to drain but if a reverse bias is applied to the base, or gate, the *depletion layer* at the p–n boundary increases in thickness, effectively narrowing the channel through which the electrons can diffuse. Note that the current flow here is controlled by a voltage rather a base current as in a simple p–n–p transistor. Since the gate voltage is in the reverse-bias mode, the input resistance of the FET is very high (100 MΩ).

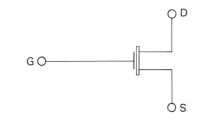

Fig. 11.14. Representation of an n-channel metal oxide semiconductor field effect transistor (MOSFET).

In the MOSFET device (Fig. 11.14) the gate is insulated from the drain and source by a metal oxide layer, enabling even higher input resistances to be attained (10^{15} Ω). Amplifiers with high input impedances are required in routine laboratory measurements of pH and in patch-clamp microelectrode measurements of the properties of individual membrane channels (Section 11.19).

11.8   Amplifiers

An 'ideal' amplifier to study small signals in systems with high associated resistances would have the following characteristics: (1) infinite gain, (2) infinite input resistance, (3) zero output resistance, (4) infinite frequency response.

Modern amplifiers that come close to these specifications are termed *operational amplifiers* (op amps) and in reality they have the following characteristics: (1) 10^6 gain, (2) 10^{10} Ω input resistance, (3) 0.1 Ω output resistance, (4) responds in the frequency range zero (DC) to 100 MHz.

Input ————⟩— Output   **Fig. 11.15.** Symbol for op amp.

Operational amplifiers

These are widely used as biological amplifiers and the symbol for the op amp is given above (Fig. 11.15). It has two input terminals (differential input) namely non-inverting (+) and inverting (−). These imply that a positive signal applied to + will give a positive output voltage. The output is proportional to the difference between the inputs. There are initially two simple rules which help in understanding the potential uses for operational amplifiers: (1) the input impedance is infinite so no current flows into the op amp; (2) the gain is infinite so that the differential input voltage is small enough to be neglected.

1 The non-inverting amplifier (Fig. 11.16)

This example shows the importance of feedback resistors in determining the gain of an op amp.

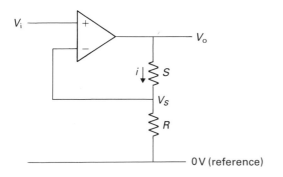

Fig. 11.16. The non-inverting op amp. The input voltage (V_i) is connected to the non-inverting input, while the inverting input is connected through a feedback resistor (S) to the output. This input is also connected to the reference voltage (usually ground) via a second resistor (R). All voltages (V_i, V_S and V_o) are measured relative to the reference.

Since no current flows through the amplifier, the current (i) through S must equal that through R. Since $V_o = i (R + S)$ and $V_s = iR$ then

$$V_S = \frac{V_0 R}{R + S}.$$

From Rule 2: $V_i = V_S$, hence $V_i = V_0 R/(R + S)$.

V_0/V_i is the amplification of the system $= \dfrac{R + S}{R}$ and for a gain of

$1000 \times$ for example, we have to set $R = 1\ \Omega$ and $S = 999\ \Omega$.

2 Differential amplifier (Fig.11.17)

This example shows how an op amp can be used to remove noise (e.g. 50 Hz from power supplies) from small input signals.

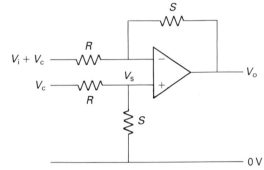

Fig. 11.17. Differential amplifier arrangement. The voltage of interest (V_i), together with the noise signal to be rejected (V_c), are connected through a resistor R to the inverting input, while the noise signal alone is connected to the other input. The non-inverting input is also connected through the resistor S to the reference voltage. The inverting input is connected to the output via the feedback resistor S.

V_i is the difference in voltage we wish to amplify.
V_c is the so-called 'common mode' voltage we wish to reject.

From Rule 1

$$V_S = V_c S/(R + S).$$

From Rule 2 The voltage on the inverting input is also equal to V_S.
From Rule 1, the current through R is equal to the through S.

Therefore

$$\frac{V_i + V_c - V_S}{R} = \frac{V_S - V_0}{S}$$

$$\therefore \qquad V_i S = V_S (R + S) - V_0 R - V_c S.$$

Substituting for V_S:

$$V_i S = -V_0 R.$$

Note that the output signal only contains the amplified differential signal V_i and not the common signal V_c.

$$\frac{V_0}{V_i} = \text{amplification}$$

$$= -S/R.$$

The output is an amplified, inverted form of the input.

Operational amplifiers can also be used to derive the integral or differential of a time-varying signal.

Op amp as a differentiator

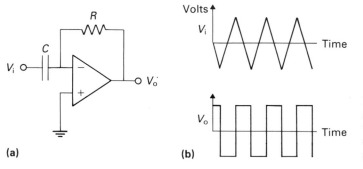

(a)   (b)

Fig. 11.18. (a) The op amp as a differentiator. (b) A 'sawtooth' waveform on the input of a differentiating op amp produces a 'square wave' output.

From Rule 1 the currents through C and R must be equal and since both input terminals must be at 0 V, the voltage across R is V_0. Hence,

$$\text{current through } C = C\frac{dV_i}{dt} = -\frac{V_0}{R}$$

$$\therefore \qquad V_0 = -RC\frac{dV_i}{dt}.$$

The output is therefore the differential of the input and the multiplying constant is RC.

Op amp as an integrator

In order to obtain integration, the resistor and capacitor are interchanged. (Fig. 11.19a)

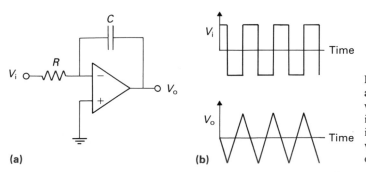

(a)   (b)

Fig. 11.19. (a) The op amp as an integrator. (b) In this case when a square waveform (V_i) is applied to the input, an integrated 'sawtooth' waveform (V_0) is obtained at the output.

From Rule 1:

the current through $C = \dfrac{\mathrm{d}Q}{\mathrm{d}t} = C\dfrac{\mathrm{d}V_0}{\mathrm{d}t} = -\dfrac{V_i}{R}.$

Hence $V_0 \quad = -\dfrac{1}{RC}\displaystyle\int V_i \mathrm{d}t.$

The output is now the integral of the input.

11.9   Two examples of op amp use

1   Measurement of ion activity (concentration)

High resistances are encountered in biology when, for example, diffusion processes are being studied. For example, the measurement of the diffusion of ions through membrane channels requires very high

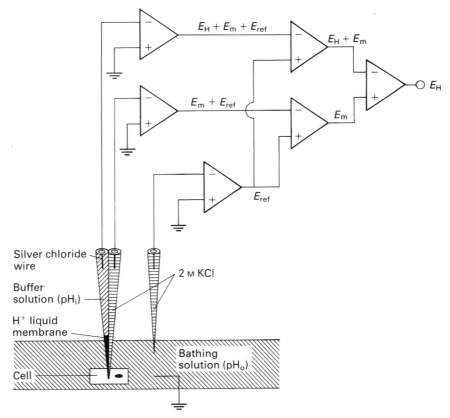

Fig. 11.20. Diagram of electrode and amplifier systems used to measure ion activity within a single cell. The final measured potential (E_H) is set up by the diffusion of protons across the liquid membrane in the tip of the electrode, i.e.

$E_H = -\dfrac{RT}{F}\ln\left[\dfrac{H_i}{H_c}\right]$ where H_i is the hydrogen ion activity in the standard buffer

solution and H_c the activity in the cell. Note that this figure is purely diagrammatic. No feedback resistors have been included and as the amplifiers invert at each stage, the final outcome is actually $-E_H$.

input impedance amplifiers (Section 10.19). A more common diffusion example arises in the measurement of pH either in laboratory solutions or in the cytoplasm of a single cell (Fig. 11.20). In both cases the hydrogen ion activity (pH) is measured relative to the activity in a standard solution. A membrane, permeable only to hydrogen ions, separates the standard solution from the solution of interest and the diffusion of H^+ across this membrane sets up a diffusion potential (Nernst potential) which can be measured (E_H). Both the glass membranes used for the measurement of pH in laboratory solutions and the liquid membrane microelectrodes employed in single cell measurements have very high resistances ($>10^{12}$ Ω in the latter case) and therefore amplifiers with very high input impedances have to be employed in the measuring circuits. In both cases the potential developed across the H^+ selective membrane has to be measured with respect to a reference electrode (E_{ref}). This electrode is simply placed in the container adjacent to the pH electrode when the solution measurement is made. In the case of the cell measurement, the reference electrode has to be placed inside the cell and for this purpose double-barrelled intracellular microelectrodes are commonly used. The tip of one barrel contains the H^+-sensitive liquid membrane and the electrode is back-filled with a standard pH buffer solution. The other barrel normally contains a high-ionic strength solution (2 M KCl) to make good electrical contact with cell cytoplasm. A silver wire is placed in both solutions to provide electrical contact to the amplifier circuit. The solution bathing the cell is normally at earth potential (E_{ref}) and the cell membrane potential (E_m) is measured with respect to a reference placed in the bath solution. The voltage measured by the H^+-sensitive electrode comprises both the H^+ voltage and the membrane voltage and the latter has to be subtracted from the composite reading on the pH electrode so that the H^+ diffusion voltage alone can be determined. This voltage can be related to cellular pH after calibrating the electrode (before and after it has been placed in the cell) in solutions of known pH.

Liquid membranes are now available that are sensitive to calcium, magnesium, sodium, potassium and chloride ions and hence this technique is very widely used to study a range of cellular mechanisms. In the example illustrated in Fig. 11.21, pH and calcium ion-sensitive electrodes are being used to study the cellular control of membrane permeability.

2   Patch-clamp methods to study membrane ion-channels

The patch-clamp technique involves pressing the heat-polished tip of a glass micropipette against the membrane of a living cell and simultaneously applying gentle suction (Fig. 11.22). The membrane–glass

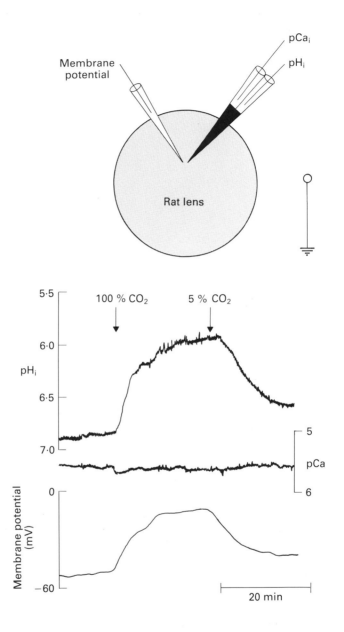

Fig. 11.21. In the experiment illustrated above, two microelectrodes were inserted into the same region of a rat lens which was continuously perfused with artificial aqueous humour solution at 35°C. The three potentials involved. (E_m, E_{pH} and E_{Ca}) were measured with respect to an earthed silver wire in the bathing medium. Both pH and calcium electrodes were calibrated with solutions of known pH and pCa (pCa $= -\log_{10}[\text{Ca}]$) before and after impaling the lens. The lens membrane potential (E_m) is depolarized on increasing the external CO_2 from 5 to 100%, and pH and pCa traces from the double-barrelled electrode demonstrate that this is due to an internal acidification and not to a change in internal calcium (data from Bassnett & Duncan, 1988).

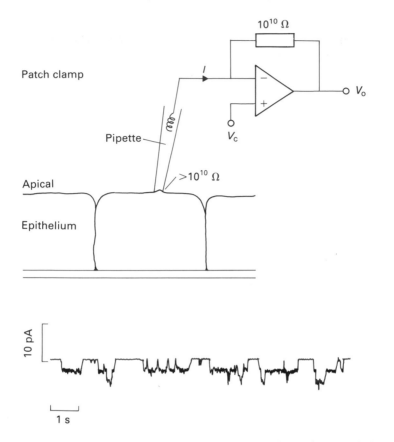

Fig. 11.22. The patch-clamp technique. In the experiment illustrated, the patch electrode has been placed against the apical surface of a lens epithelial cell and, with the potential across the patch clamped at the resting potential, fluctuations in the clamp current are observed. These currents are due in this instance to the flow of Na^+ into the cell, although K^+ and Ca^{2+} can also flow through this non-specific cation channel. It is believed that this channel plays a role in the development of cortical cataract. Note that more than one channel is present in the patch as the unitary current events show evidence of summation on a number of occasions (data from Jacob *et al.*, 1985).

interaction produces seals with a leak resistance to the bath in the range 10–100 G Ω. This high resistance ensures that currents flowing through the patch of membrane isolated under the tip of the electrode flow into the pipette. The internal filling solution of the microelectrode is connected via a silver wire to the input of a high impedance amplifier (op amp). The other input of the amplifier is connected to an external voltage source (V_c) so that the potential of the membrane patch clamped at a value that is under the control of the experimenter. Since no current can pass through the amplifier, the output V_0 is a measure of the single channel currents I passing through the feedback resistance R (10^{10} Ω) where $I = V_0/R$.

A very large number of channels have been investigated by patch-clamp techniques and they include the voltage-gated sodium and potassium channels involved in generating nerve and muscle action potentials as well as the ligand-gated channels involved in various types of neurotransmitter and hormone actions. Some channels, however, do not have to be activated and are present at the resting potential of the cell. An example of this type of channel is the non-selective cation channel present in many cell types. This channel does not discriminate between sodium and potassium (Fig. 11.22).

By careful manipulation it is also possible to tear off the isolated membrane patch from the cell (excised patch). This method was used to great effect in the study of the gating mechanism of rod photoreceptor

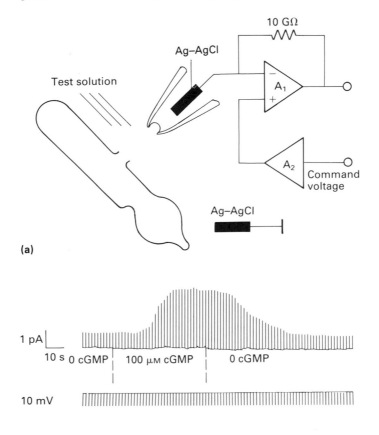

(a)

(b)

Fig. 11.23. The effect of cyclic GMP on a patch of plasma membrane excised from a rod outer segment. Constant command voltage pulses were applied to the patch (lower trace), and the currents produced by them were measured by the op amp A_1 (upper trace); the currents are thus proportional to the membrane conductance. When cyclic GMP was applied to the cytoplasmic side of the patch the conductance increased fourfold, indicating the presence of channels that are held open by cyclic GMP (from Fesenko *et al.*, 1985).

channels (Fig. 11.23). The total conductance of the patch was measured in this instance, and it was found that addition of cyclic GMP to the solution facing the cytoplasic face of the membrane increased the conductance fourfold. Calcium did not modulate this conductance directly and this was the first experiment that helped resolve whether Ca^{2+} or cGMP gated rod photoreceptor sodium channels. The role which calcium has to play is nevertheless important as Ca^{2+} influences the phosphodiesterase system involved in cGMP hydrolysis.

The cGMP gated channels in the rod plasma membrane do not in fact discriminate well between Na^+ and K^+ and they also allow Ca^{2+} to pass. They are probably related to the non-specific cation channels described above.

References

Bassnett, S. & Duncan, G. (1988) The influence of pH on membrane conductance and intercellular resistance in the rat lens. *Journal of Physiology,* **398**, 507–21.

Fesenko, E.E., Kolesnikov, S.S. & Lyubarsky, A.A. (1985) Induction by cyclic GMP of cationic conductance in plasma membrane of retinal rod outer segment. *Nature,* **313**, 310–13.

Jacob, T.J.C., Bangham, J.A.B. & Duncan, G. (1985) Characterisation of a cation channel on the apical surface of the frog lens epithelium. *Quarterly Journal of Experimental Physiology,* **70**, 403–21.

Further reading

Brown, B.H. & Smallwood, R.H. (1981) *Medical Physics and Physiological Measurement.* Blackwell Scientific Publications, Oxford.

Delaney, C.F.G. (1969) *Electronics for the Physicist.* Penguin, Middlesex.

Faulkenberry, L.M. (1977) *An Introduction to Operational Amplifiers.* Wiley, New York.

Hamill, O.P., Marty, A., Neher, E., Sakmann, B. & Sigworth, F.J. (1981) Improved patch-clamp techniques for high-resolution current recording from cells and cell-free membrane patches. *Pflügers Archives,* **391**, 85–100.

Morant, M.J. (1964) *Introduction to Semiconductor Devices.* Harrap, London.

Rink, T.J., Tsien, R.Y. & Warner, A.E. (1980) Free calcium in *Xenopus* embryos measured with ion-selective microelectrodes. *Nature,* **283**, 658–60.

Sakman, B. & Meher, E. (1983) *Single-Channel Recording.* Plenum Press, New York.

Standen, N.B., Gray, P.T.A. & Whitaker, M.J. (1987) *Microelectrode Techniques.* Company of Biologists, Canbridge, U.K.

12 Magnetism and Electromagnetism

12.1 Introduction

Magnetism is an area of physics that appears at present to have only a limited role in biological mechanisms but is of immense importance in modern instruments used to investigate a vast range of biological and medical phenomena. For example, magnetic field detectors *may* play a role in bird navigation and application of magnetic fields *appear* to promote bone and tissue healing but there is little doubt that the technique which has the greatest potential for imaging cells and tissues in the living state is that of *nuclear magnetic resonance* (NMR). It is possible with this technique not only to produce images from single cells

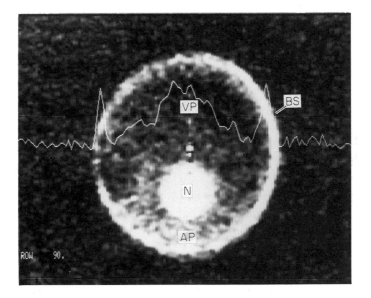

Fig. 12.1. NMR image representing a transverse slice across a glass tube containing a stage 4 *Xenopus laevis* ovum: it has a spatial resolution of 10×13 μm and a slice width of 250 μm. The continuous profile demonstrates the good signal-to-noise ratio in the nuclear region (N). The data used to form the image arise from proton nuclear magnetic moments resonating under the influence of intense magnetic fields. The different regions seen correspond to differences in the state (relaxation times) of water in the cell. Interestingly, nuclear water seems less strongly bound to proteins than cytoplasmic water as its signal (light region) has similar relaxation characteristics to water in the adhering bathing solution (BS). The animal (AP) and vegetal (VP) poles also show different characteristics. This image is extremely important as it was the first to give spatial and chemical information from a single cell by NMR techniques (from Aguayo *et al.*, 1986).

(Fig. 12.1) and whole organs (Section 12.11), but also to investigate molecular interactions in specific areas of these cells and organs.

12.2   Magnetic domains and fields

A magnetized bar behaves as if it were asymmetrical in some way — it is said to have two poles, one *north* and the other *south*. Since two magnets are produced every time a magnetized bar is cut, the bar can be visualized as being composed of very small *magnetic domains*. In an unmagnetized bar these would take up a random orientation while in a magnetized bar they would be lined up (Fig. 12.2). An iron bar can be magnetized by 'stroking' it in one direction with another magnet or by placing it inside a coil of wire through which a current is flowing.

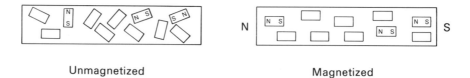

Unmagnetized   Magnetized

Fig. 12.2. The positions of the north and south poles in the hypothetical magnetic domains are denoted N and S respectively.

When the two north poles from two magnets are brought together the magnets repel one another. A magnet is said to have a magnetic field associated with it which modifies in some way the space surrounding the magnet. This field will interact with the field from another magnet if the two are brought together and a repulsive or attractive force will result.

The magnetic field of a bar magnet is shown in Fig. 12.3. The density of the field lines, the *magnetic flux density*, is a measure of the intensity of the magnetic field. Soft iron bars tend to concentrate field lines

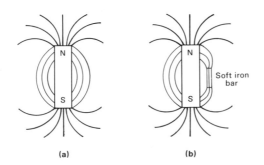

(a)   (b)

Fig. 12.3. See text for explanation.

(Fig. 12.3b) so the flux density inside the iron bar is greater than it would be if the bar was replaced by air or glass.

12.3   Electrical currents and magnetic fields

Magnetic fields are also set up by currents flowing along conductors and this can be clearly seen if a compass needle is placed near a wire carrying a direct current (Fig. 12.4).

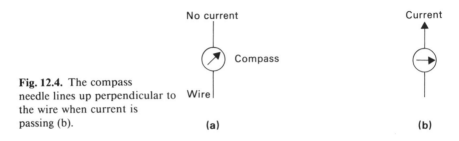

No current   Current

Compass

Fig. 12.4. The compass needle lines up perpendicular to   Wire the wire when current is passing (b).
                                                 (a)   **(b)**

Two wires a short distance apart exert forces on one another as a result of the interaction of the fields (Fig. 12.5). If the currents are flowing in the same direction the force is an attractive one and it is repulsive if the currents are moving in opposite directions. The forces depend on the currents flowing in the wires and their distances apart and, as the force can be very easily measured, the arrangement is used to define the *ampère*.

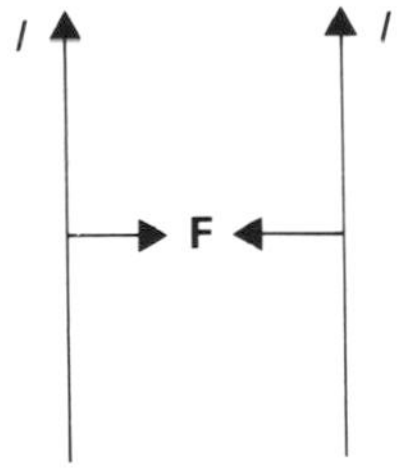

Fig. 12.5. Two wires with current flowing in the same direction experience an attractive force **F**.

The ampère (amp) is that intensity of current which, flowing in each of two long parallel conductors 1 m apart, in a vacuum, results in a force of 2×10^{-7} N m^{-1} between the conductors. This also serves to define the *coulomb* as the charge transferred by a current of one ampère flowing for one second.

Magnetic fields exert forces on current-carrying conductors (Fig. 12.6) and the vector relationship between the two is given by

$$\mathbf{F} = IL \times \mathbf{B} \tag{12.1}$$

where **B** is the magnetic field; I is the current in the conductor, and L is the length of the conductor. This equation serves as a definition of the units of magnetic field. One *weber* m^{-2} (or one tesla, T) is defined as that magnetic field which will exert as a force of 1 N on a 1 m length of

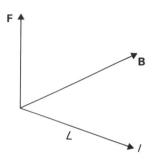

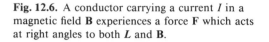

Fig. 12.6. A conductor carrying a current I in a magnetic field **B** experiences a force **F** which acts at right angles to both L and **B**.

wire carrying a current of 1 A and lying in a direction at right angles to the magnetic field. The converse phenomenon is also true. If a conductor is moved in a magnetic field the charges within it experience a force. The movement of these charges constitutes a flow of current which in turn produces a force tending to oppose the movement of the conductor. Work will then have to be done to move the conductor. Suppose the conductor moves a small distance dx, then the work done, dW, is given by

$$\mathrm{d}W = \mathbf{F} \cdot \mathrm{d}x = IL \times \mathbf{B} \cdot \mathrm{d}x. \tag{12.2}$$

The power input is dW/dt, and therefore

$$\text{power} = IL \times \mathbf{B} . \mathbf{v} \tag{12.3}$$

where **v** is the velocity of the conductor. But the power input is also given by IV (equation 10.12), therefore

$$V = L \times \mathbf{B} . \mathbf{v}. \tag{12.4}$$

V is the pd across the ends of the conductor and **v** is the velocity of the conductor. **B** is the flux density of magnetic field in Wb m^{-2}.

When L, **B** and **v** are mutually perpendicular, as in the case of the electromagnetic flow meter (Section 12.4), equation (12.4) simplifies to

$$V = L \, B \, v. \tag{12.5}$$

Electromagnetic flow meter

The electromagnetic blood flow transducer (Fig. 12.7) consists of an electromagnet (see following section) to generate a magnetic field and two electrodes to sense the voltage across the conductor (V_f). They are encapsulated in an inert hard plastic in a form which permits them to fit around the blood vessel of interest. The lumen or inside diameter of the holder slightly deforms the vessel so that its cross-sectional area is now fixed and indeed known. In this way the transducer can be used to measure the flow J although basically it is a mean velocity transducer (i.e. $V_f \, \alpha \mathbf{v}$).

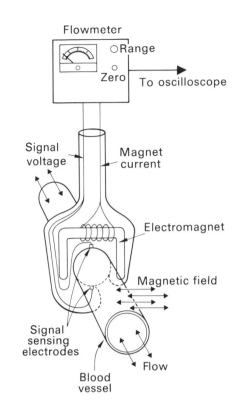

Fig. 12.7. Electromagnetic flowmeter. Equation (12.5) can be applied directly to this system as the blood flow, i.e. velocity of the conductor, the conductor length (distance between sensing electrodes), and the magnetic field are all at right angles to one another (from Strong, 1970). (Reproduced by permission of Tektronix Inc.)

Problem 12.1

An electromagnetic flowmeter with a sensing head of cross-sectional area 0.07×10^{-4} m^2 and a fixed distance between sensing electrodes of 1×10^{-3} m is used to measure blood flow in an artery. When a magnetic field of 2 Wb m^{-2} was applied, a peak signal of 600 μV was observed on the oscilloscope screen. Calculate the blood flow (m^3 s^{-1}) through the artery.

12.4   Electromagnetism

When a conducting wire is wound in the form of a helix very strong magnetic fields are produced along the axis when current flows along the wire (Fig. 12.8). The direction of the field depends on whether the current (I) flows in a clockwise or anticlockwise direction and the magnitude of the induced field (B) is given by:

$$B = \frac{\mu_0 NI}{l} \qquad (12.6)$$

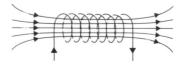

Fig. 12.8. The magnetic field of a coil.

where μ_0 is called the *permeability of free space* and has the value $4\pi \times 10^{-7}$ Wb A^{-1} m^{-1}; N is the number of turns of the helix which is of length l. If iron or some other *ferromagnetic* material is inserted into the helix the field increases to:

$$B = \frac{\mu_r \mu_0 N I}{l}$$

where μ_r is the *relative permeability* of the ferromagnetic material. In the case of iron this has the value 5000, so intense fields can be produced. When the iron is bent into the form of a circle (Fig. 12.9), strong, homogeneous fields are set up across the ends and this arrangement can be used to produce the very high field strengths necessary in nuclear magnetic resonance spectroscopy.

Fig. 12.9. An electromagnet.

12.5   Magnetic resonance spectroscopy

Magnetic fields exert an influence on atoms and molecules both at the electronic and nuclear levels. In electron paramagnetic resonance spectroscopy (EPR, or electron spin resonance, ESR) the magnetic moments resulting from unpaired electrons can be aligned in a strong magnetic field (≈ 6.3 T) and transitions between energy levels can occur when the oriented moments absorb energy in the microwave range ($\approx 10^{10}$ Hz). Most biological molecules in fact have paired electrons and so do not interact in this way, but unpaired electrons do exist in some transition metal ions and in free radicals. Stable free radicals (e.g. nitroxides) can be manufactured and attached to biologically interesting molecules. The biological and medical applications of EPR include the detection of harmful free radicals, the study of the ligand environment around a metal site in a metalloprotein and the measurement of molecular motion in membranes, for example, which is difficult to measure by other techniques.

In nuclear magnetic resonance (NMR) nuclear magnetic moments can be aligned in very strong magnetic fields and resonance transfer of energy can occur when energy is supplied in the radiofrequency range (≈ 100 MHz). Since this latter technique, which is in principle similar to EPR, is very widely used in biology and medicine it will be described in some detail.

12.6 Nuclear magnetic resonance

Nuclear magnetic resonance techniques have been applied to biological and medical problems with great success over the past 10 years and, because of the versatility of this technique, there is little doubt that major advances will be made at the research and clinical level during the next decade. NMR can be used to obtain information about the structure of individual molecules and their interaction with the environment but, more importantly, it can now be used to obtain images of the structure of individual cells (Fig. 12.1) as well as parts of the human body. In the future it will no doubt be used to obtain three-dimensional images of chemical reactions and molecular interactions within single cells, tissues or even whole animals.

The physical principles of operation are the same whether pure chemicals or whole bodies are being investigated. The technique depends on the fact that certain chemical species (e.g. the abundant proton) have a *net nuclear magnetic moment*. This moment can be manipulated in intense magnetic fields so that energy is absorbed and re-emitted (Fig. 12.10). This energy exchange depends not only on the type of nuclear moment involved, but also on its chemical environment and its interaction with surrounding nuclei.

The nuclear magnetic moment arises from the fact that the nucleus is spinning (Fig. 12.11). This moment can interact with a magnetic field

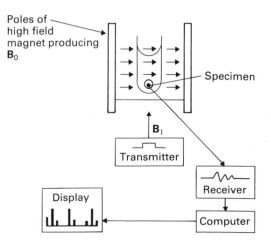

Fig. 12.10. Block diagram of a nuclear magnetic resonance spectrometer. The specimen is placed between the poles of a magnet producing an intense field ($\mathbf{B}_0$). A pulse of radiowaves ($\mathbf{B}_1$) from the transmitter causes nuclei in the sample to change their orientation and the resultant signal is detected by the radiofrequency receiver and stored in the computer. After processing (usually by Fourier-transform methods) the results are displayed and plotted.

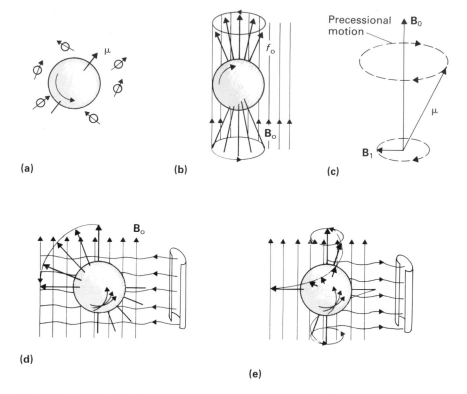

(a)   (b)   (c)

(d)

(e)

Fig. 12.11. (a) Nuclei with net nuclear moment (μ) can be considered to be spinning like tops, and in any tissue or solution the moments will be randomly aligned. (b) The effect of an external magnetic field ($\mathbf{B}_0$) is to cause the magnetic moments to line up and precess around the direction of the field. The precessional frequency (f_0) increases with field strength. (c) Diagram showing the interaction between the high field $\mathbf{B}_0$, the nuclear magnetic moment μ and the transmitter field $\mathbf{B}_1$. The magnetic moments (μ) precess about the field $\mathbf{B}_0$ and, when the rotating field supplied by the transmitter and the precession have the same frequency, transfer of energy can occur and the nuclear magnet will change its orientation. (d) If a precessing nucleus is subjected to a radiofrequency field of the same frequency as the precession (f_0), it is knocked out of alignment. Within seconds it spirals back to the original minimal energy orientation and in doing so it emits a burst of energy in the radiofrequency range and this is detected by the receiver (e).

and the potential energy of the interaction is a minimum when the moment is pointing along the field, just as a compass takes up a position of minimal potential energy in the earth's magnetic field by pointing north–south, rather than south–north. The energy (E) of the interaction depends on the projection of the moment along the field,

$$E = -\mu.\mathbf{B}_0$$

where μ is the moment and $\mathbf{B}_0$ the field (Fig. 12.11c).

In reality, the magnetic moment does not lie precisely along the magnetic field, but at some angle to it and this gives rise to a permanent torque. The nucleus has angular momentum because of its spin and the net result of the interaction of angular momentum and torque is *precession*. This outcome can clearly be seen in the movement of a child's spinning top which precesses while it spins in the earth's gravitational field. The frequency of precession of the nucleus is given by

$$\omega_0 = \gamma \mathbf{B}_0$$

where $\omega_0 (= 2\pi f_0)$ is called the resonance or Larmor frequency, and γ is a proportionality constant that is different for each nucleus. If energy is supplied externally via a radio frequency oscillator then energy exchange can occur and the nuclear magnet will change its orientation from the low to the high energy state.

The nucleus will, in a short time, emit this energy as it returns to the low energy state and this process is detected by the radiofrequency receiver (Fig. 12.10). Theoretically it is possible to excite the sample with a range of radiofrequencies and to tune the receiver to detect the appropriate frequency. It is, however, much simpler to transmit a pulse of radiofrequency energy to the sample. The transmitter is turned off and the receiver on (Fig. 12.12). The emitted signal consists of complex waveform depending on the relaxation times of the nuclei involved and the relative amplitude of their contribution to the signal. This amplitude information in the 'time domain', called the free induction decay signal, has to be converted to amplitude information in the 'frequency domain' (Fig. 12.12). The mathematical process which achieves this transformation, and which is applicable to a wide range of processes, is called Fourier transformation (see Section 7.3). There are a number of

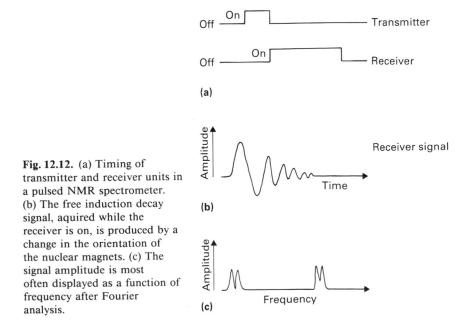

Fig. 12.12. (a) Timing of transmitter and receiver units in a pulsed NMR spectrometer. (b) The free induction decay signal, aquired while the receiver is on, is produced by a change in the orientation of the nuclear magnets. (c) The signal amplitude is most often displayed as a function of frequency after Fourier analysis.

physical processes that determine (a) the frequency at which the absorption peaks occur in the spectrum, (b) the shape of the absorption peak and (c) the number of peaks associated with any one magnetic nucleus.

12.7   Chemical shift in NMR spectrum

The electronic environment of the nucleus provides a magnetic shield (Fig. 12.13). If the nucleus is surrounded by a symmetric cloud of electrons, the applied field will cause the electrons to circulate, producing an induced magnetic field ($\mathbf{B}_{ind}$) which opposes $\mathbf{B}_0$ and is proportional to it.

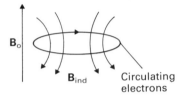

Fig. 12.13. The applied field $\mathbf{B}_0$ can be reduced in nuclei that are surrounded by a symmetric electron cloud. The induced field ($\mathbf{B}_{ind}$) is opposite in direction, and proportional to, the applied field.

$$\mathbf{B}_{ind} = \sigma\mathbf{B}_0.$$

The field experienced by the nucleus is given by

$$\mathbf{B}_{local} = \mathbf{B}_0\,(1 - \sigma),$$

where σ is termed the *shielding constant*. This local field effect can be used to characterize certain groups. For example the proton resonance signal associated with the OH moiety in ethanol occurs at a different frequency from those associated with the CH_2 and CH_3 groups (Fig. 12.15). Instead of presenting the data directly in terms of frequency, which would be meaningless unless the field strength of the magnet was also given, they are given as the chemical shift (δ) relative to the resonance frequency (v) of some standard, i.e.

$$\delta = \left(\frac{v_{standard} - v_{sample}}{v_{standard}}\right) \times 10^6.$$

For protons, the standard most often used is tetramethylsilane (TMS): $(CH_3)_4$–Si and Fig. 12.14 gives some approximate chemical shifts for main proton groups. (Values for δ are usually quoted in parts per million (ppm).)

12.8   Spin–spin coupling

Close inspection of the ethanol spectrum (Fig. 12.15) reveals that only the proton peak associated with the OH group is a single peak. The other peaks have split because of spin–spin coupling between adjacent

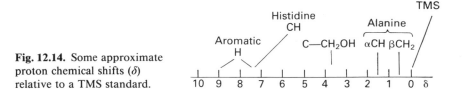

Fig. 12.14. Some approximate proton chemical shifts (δ) relative to a TMS standard.

nuclear spins and these interactions are communicated between the nuclei by electrons in a chemical bond. The electron cloud communicates the direction of one nucleus to the other because it becomes slightly polarized by the nuclear spin. The relative intensities and number of lines in the multiplet splitting depends on the number of ways of orienting the spins responsible for the splitting (Fig. 12.16).

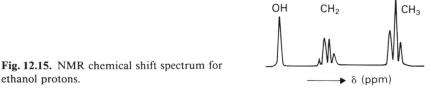

Fig. 12.15. NMR chemical shift spectrum for ethanol protons.

The resonance from the CH_2 group is a quartet because it is adjacent to three identical protons in the CH_3 group, while the resonance from the CH_3 group forms a triplet as it is adjacent to the two identical protons from the CH_2 group. The coupling between the OH hydrogen and the CH_2 group is removed because of rapid chemical exchange of this hydrogen between different ethanol molecules.

The ^{31}P spectrum from ATP (Fig. 12.17) shows three major resonances corresponding to the α, β and γ phosphorus atoms. The α and γ peaks are doublets because of their interactions with the β phosphorus, while the resonance from the β forms a triplet because of the interactions with the α and γ atoms.

Note that the magnitude of the splitting is termed the 'coupling constant' for the interaction and is denoted by the symbol J. The field produced at nucleus (1) by nucleus (2) is independent of B_0, so that J, measured in hertz has the same value for any given pair of nuclei at any value of the measuring frequency.

12.9 Two-dimensional NMR

A limitation of conventional NMR is that overlap of resonance signals from two different processes can occur. This is avoided by using a double pulse technique so that there are two time variables (t_1 and t_2) rather than just one. The two pulses are applied during t_1, before

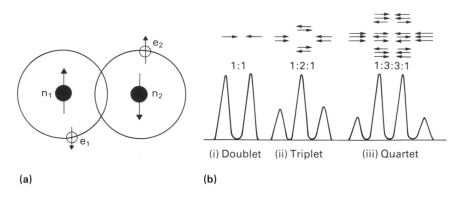

(a)

(b)

Fig. 12.16. (a) Electron-mediated spin–spin coupling. Each electron tends to be paired with its nucleus and with the other electron. The result at n_2 is that the local field will depend on the orientation of n_1. If n_2 is adjacent to a single nucleus, there are two possible orientations for n_1 (spin up and spin down). The spin–spin spectrum will therefore comprise two peaks of equal magnitude. When n_2 is adjacent to two identical nuclei, there are four possible spin coupling orientations (two of them identical), so a triplet is obtained. With three adjacent nuclei, a quartet is obtained. (b) Splitting in NMR spectra due to spin–spin coupling observed when a neighbouring group has (i) one spin, (ii) two equivalent spins, (iii) three equivalent spins. Top — possible orientations of neighbouring group (after Knowles *et al.*, 1976).

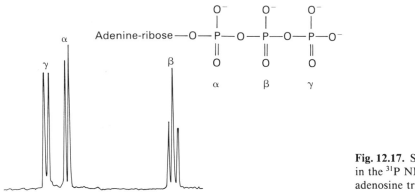

Fig. 12.17. Spin–spin splitting in the ^{31}P NMR spectrum of adenosine triphosphate.

sampling during t_2 (Fig. 12.18). Many spectra are then acquired in which the t_1 values are changed. A Fourier transformation is then performed with t_2 as the time variable to give amplitude information in one dimension (Fig. 12.19), followed by a second Fourier transformation with t_1 as the variable to give the other dimension. By changing t_1 either more or less time is allowed for evolution due to spin–spin coupling which cannot be refocused with the 180° pulse. Hence f_1 will contain information only concerning spin–spin coupling resonances while f_2 will relate to chemical shifts alone. The proton spectrum obtained from xylose shows this very clearly (Fig. 12.19). By using different magnetic combinations of pulses during t_1 it is also possible, for example, to obtain spectra with ^{13}C chemical shifts along one axis

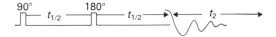

Fig. 12.18. Double pulse technique to obtain NMR data containing two sets of information. The second pulse is often applied mid-way through t_1 and, in the case illustrated in Fig. 12.19, it is applied with a 90° phase shift relative to the first pulse.

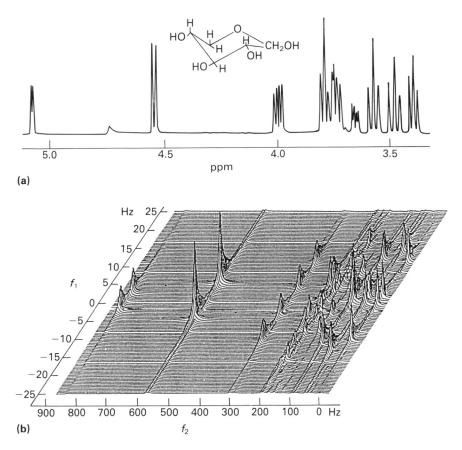

(a)

(b)

Fig. 12.19. (a) ^1H spectrum of xylose D_2O solution taken at 400 MHz. (b) *J*-resolved two-dimensional spectrum of xylose taken at 400 MHz; f_1 axis 50 Hz, 12 points, f_2 axis 900 Hz, 2048 points; 16 acquisitions (by permission of Joel Co. Ltd.).

and ^1H shifts along the other. A very large number of combinations are possible and the NMR manufacturers' 'application notes' are an excellent source of information.

12.10 Solid state NMR

NMR resonance peaks are not infinitely sharp and the extent of 'line-broading' depends on the interaction of the nuclear spin with the environment with associated local differences in magnetic field. In rapid motion these fields average out but in solids a broad band is seen (Fig. 12.20a). The broadening in chemical shift resonances can be largely eliminated by rapidly spinning the solid sample at an angle of 54.7° with the magnetic field B_0. This angle (the so-called 'magic angle') causes the function $1-3\cos^2\theta$ to vanish (this function appears in a mathematical analysis of sample-spinning experiments). Even greater sharpening up of the spectra can be obtained by using a pre-pulse technique in combination with magic angle spinning to manipulate the magnetic moments. The spectrum of β-quinol-methanol clathrate obtained by spinning at 54.7° (Fig. 12.20) explains why it is termed 'magic'.

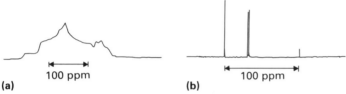

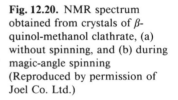

Fig. 12.20. NMR spectrum obtained from crystals of β-quinol-methanol clathrate, (a) without spinning, and (b) during magic-angle spinning (Reproduced by permission of Joel Co. Ltd.)

100 ppm

(a)

100 ppm

(b)

12.11 NMR imaging

One of the most exciting aspects of NMR is the ability to obtain chemical images from cells, tissues and even whole bodies using this technique. Two-dimensional information can be obtained from a specimen by applying two gradients of magnetic fields at right angles to one another. The very first published two-dimensional NMR 'picture' was of two tubes of water (A and B in Fig. 12.21).

When the field is applied horizontally (G_x), two signals are obtained, while only one is obtained when a vertical field is applied (G_y). When a combination of G_x and G_y is used, the effective viewing angle depends on the relative strength of G_x and G_y ($G_y + G_y$). The direction and thickness of the slice itself can be chosen by applying a gradient in the Z direction (G_z).

Most modern imaging systems, whether of a single cell or of the human body, employ a double pulse method to give both spatial and chemical information.

The main DC or B_0 magnetic field is along the length of the patient (the Z axis in Fig. 12.22) and the very high field strengths required (0.15 T) are developed by cryogenic magnets. This gives a proton resonant frequency of 6.5 MHz.

The protons are excited by radiofrequency radiation transmitted by a coil surrounding the region of interest and the signals produced by the excited nuclei are detected by a receiver coil within the transmitter coil.

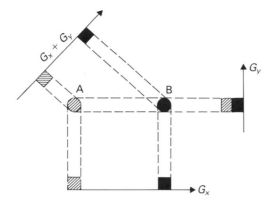

Fig. 12.21. Diagrammatic representation of the effect of linear magnetic field gradients G_x and G_y on the NMR responses of two tubes (A and B) containing water. When G_x is on, the signals from the two tubes are spread along the horizontal axis and they are clearly resolved. When the field gradient is in the y direction only one signal is obtained and when a combination of linear gradients $(G_x + G_y)$ is used the effective angle from which the sample is viewed depends on the relative strengths of G_x and G_y.

Protons are lined up by the $\mathbf{B}_0$ field, producing a net magnetization in the Z direction (Fig. 12.22). A short radiofrequency pulse $\mathbf{B}_i$ is applied to rotate the net magnetization through 180° into the Z direction. If permitted, the magnetization returns exponentially to the original direction with a time constant τ. A pulse half the length of the first, for example, will rotate the magnetization through 90° and into the XY plane. Precessional radiation emitted in this plane is detected by the receiver coil.

Slice selection in the Z axis is achieved by applying a pulsed gradient magnetic field along the Z axis at the same time as the as the 90° pulse is applied (Fig. 12.23). Only protons within the slice will resonate with the applied 90° pulse. The X–Y coordinates of the excited protons in the slice are obtained by applying a small vector gradient of field in the XY plane and during this time the precessional motion in this plane generates a signal which is detected by the receiver coil. Each pulse sequence thus gives one projection of the image and, for successive projections, the vector gradient is rotated 1° at a time to 180°. Image reconstruction is achieved using Fourier transformation from the frequency to the spatial domain. Protons that relax quickly after the first 180° pulse produce a large positive result by this method as they are in position for the second applied pulse; protons that relax more slowly produce a much smaller signal. On the image of a section through the human head (Fig. 12.24a) tissues such as subcutaneous fat and yellow bone marrow which have short relaxation times are shown towards the white end of the grey scale, while tissues or fluids with longer relaxation lines such as cerebrospinal fluid and ocular fluids are shown at the black

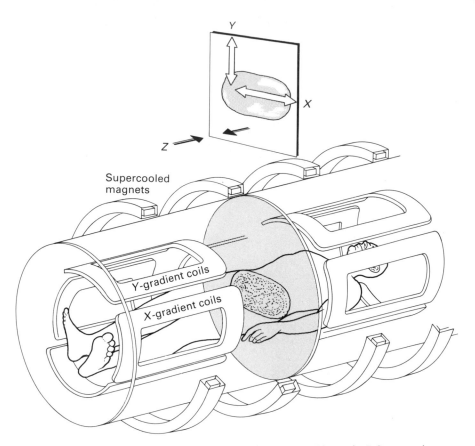

Fig. 12.22. Diagram of whole-body NMR imaging system. The main DC magnetic field (B_0) is in the longitudinal axis of the body or Z direction. The radiofrequency field (B_1) is in the XY plane and the coils located around the region of interest have been omitted for the sake of clarity. Smaller coils are positioned so that a magnetic field gradient can be applied in the Z direction to define the slice of interest. The X and Y fields are applied simultaneously to give a vector gradient in the XY plane in order to define the position of the resonating protons in this plane (after Sochurek, 1987).

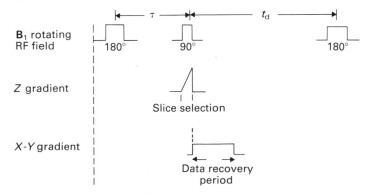

Fig. 12.23. Representation of double pulse sequence commonly used to obtain NMR imaging data. The initial 180° B_1 pulse is followed at time τ by a 90° pulse. At this stage a Z gradient is applied to provide slice selection along the Z axis. This is followed by a vector gradient in the XY plane during the data recovery period. The cycle is repeated at time t_d after the 90° pulse with the vector gradient rotated through 1°, (after Doyle *et al.*, 1981).

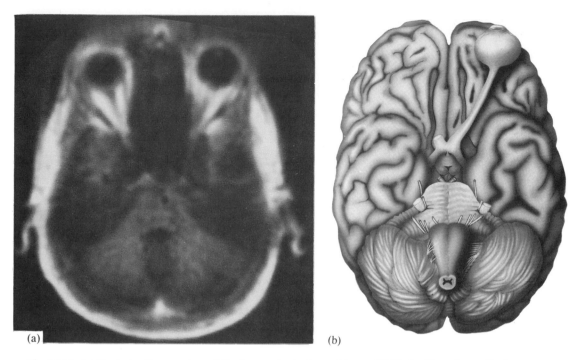

Fig. 12.24. (a) Posterior fossa region of the human brain imaged by an NMR double-pulse method (Fig. 12.23). The fourth ventricle and surrounding tissues, including the middle cerebellar peduncle and vermis are clearly resolved (from Doyle *et al.*, 1981). (b) Artist's impression of the base of the brain with left eyeball and optic nerve attached (from *Gray's Anatomy*, 30th edition).

end. Tissues with a low concentration of hydrogen such as cortical bone also appear black. A comparison with the drawing of the human head from *Gray's Anatomy* (1901) shows that image formation by NMR techniques is fast approaching the resolution obtained by the skilful draughtmanship of the early 20th century anatomists!

The power of modern NMR techniques can clearly be seen in Fig. 12.25 where four brain images from twin sisters are displayed. The girls were born at 29 weeks and examined with NMR at 9 and 33 months of age. Fig. 12.25a shows the expected level of myelination (light regions) for an infant of 9 months, while her sister, who suffered an intra- and periventricular haemorrhage at birth, shows hydrocephalus and a delay in white matter development (c). Figs 12.25b and d, obtained at 33 months show an increase in the level of myelination but the second twin remains delayed in myelination compared with her normal sister.

References

Doyle, F.H., Gore, J.C., Pennock, J.M. *et al.* (1981) Imaging of the brain by nuclear magnetic resonance. *Lancet*, **ii**, 53–7.
Joel Co. Ltd. (1981) *Application Notes: Focus on FT NMR.*

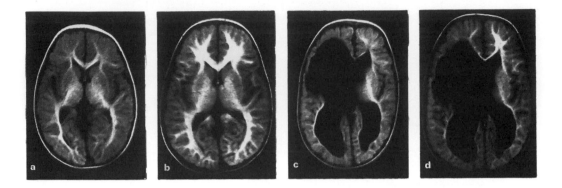

Fig. 12.25. NMR images from the ventricular region of the brain obtained by a double-pulse technique; (a) and (b) are from a normal child at 9 and 33 months of age respectively, while (c) and (d) are comparable images from her twin sister who suffered a brain haemorrhage at birth. (Photographs kindly supplied by Dr Jackie Pennock).

Knowles, P.F., Marsh, D. & Rattle, H.W.E. (1976) *Magnetic Resonance of Biomolecules.* Wiley, London.

Sochurek, H. (1987) Medicine's new version. *National Geographic*, **171**, 2–40.

Strong, P. (1970) *Biophysical Measurements.* Tektronix, Beaverton.

Further reading

Aguayo, J.B., Blackband, S.J., Schoeniger, J. *et al.* (1986) Nuclear magnetic resonance imaging of a single cell. *Nature*, **322**,190–1.

Campbell, I.D. & Dwek, R.A. (1984) *Biological Spectroscopy.* Benjamin Cummings Inc., Menlo Park, USA.

Dendy, P.P. & Heaton, B. (1987) *Physics for Radiographers.* Blackwell Scientific Publications, Oxford.

Gadian, D.G. (1982) *NMR of Living Systems.* Oxford University Press, Oxford.

Hall, L. (1986) Interrogation of molecules or man. *Laboratory Science and Technology*, **2**, 10–14.

13 Radiations and Radiobiology

13.1 Introduction

Two discoveries made at the end of the 19th century in the field of atomic physics have profoundly influenced the progress of modern biology. In 1895, Röntgen, while working with an early form of the cathode ray tube, found that rays from it caused fogging of photographic plates and he suggested that unknown, invisible radiations or 'X-rays' were being produced by the tube. He also took the first X-ray photograph by placing his hand between the tube and a photographic plate.

In the following year Becquerel discovered *natural radioactivity* and showed that uranium ores placed near photographic plates caused fogging in the dark. He was unfortunate to be one of the first scientists to suffer from radiation as he used to carry his ore samples around in his pocket.

We have previously seen that X-rays are highly energetic photons released from excited atoms during electronic transitions between the innermost orbitals (Chapter 9); natural radioactivity on the other hand is a property of the atomic nucleus.

13.2 The atomic nucleus and the radiation spectrum

The nucleus consists of nucleons, positively charged *protons* and uncharged *neutrons* held together by very strong nuclear forces set up by the exchange of elementary particles called *mesons* among the neutrons and protons. In so-called *radioactive* nuclei the forces are unable to overcome the mutual repulsion of the nucleons, and the stable state is reached by a shedding of excess energy in the form of ionizing radiations. Radioactive atoms occur naturally, but they can also be produced artificially.

The first radiations to be characterized were the α and β-particles, and γ-rays from uranium ores. γ-rays are highly energetic photons arising from a rearrangement of energy levels within the nucleus; β-particles are highly energetic electrons arising from a neutron to proton transition within the nucleus

$$n^0 \rightarrow p^+ + \beta^- .$$

(13.1)

An additional elementary particle called the antineutrino should also appear on the right side of equation (13.1) but it has been omitted as it has no known biological effect.

Some radioactive atoms, e.g. uranium, emit the very large α-particle to reach a more stable state and in this case two protons and two neutrons are emitted as one entity. *Positrons* (positive electrons), *neutrons* and *protons* can also be produced from radioactive atoms. This chapter will deal with the effects of the radiations most likely to be encountered by the biologist. The biological effects of the multitude of elementary particles found only in the depth of space or at the centre of a nuclear reactor or particle accelerator will be left to the enthusiast to research (See Thornburn, 1972).

Radiations are partly characterized by the energy of the particle or photon emitted and this information enables anyone handling a radioactive source to take adequate precautions to shield it, because the range of the radiation depends on the energy. The data for energy and range in tissue for radiations that a biologist might encounter are given in Table 13.1. The energies are given in terms of electron volts (eV).

Radiation	Energy range	Range in tissue
α-particle	1–20 MeV	1–10 μm
β-particle	10 keV–15 MeV	10 μm–5 × 10^{-2}m
X-and γ photons	10 keV–2 MeV	10^{-3}–10^{-1}m

Table 13.1. The range for the photons is the so-called half-thickness of tissue, i.e. the thickness that will reduce the initial intensity by one-half. The range of the particles is the thickness that will stop the particle. 1 keV = 10^3 eV and 1 MeV = 10^6 eV.

13.3   Sources of radiation

Anyone who has had an X-ray taken in a hospital or dentist's surgery will have been knowingly exposed to a certain amount of radiation. However, there are more, less obvious sources of radiations that are to be found all around us, and indeed in the very matter we are made of. These ubiquitous sources are the so-called *radioisotopes*.

Isotopes are atomic species that have the same number of protons in their nuclei, but different numbers of neutrons. They have the same number of electrons so their chemical properties are identical but because of the differing proportions of neutrons and protons, some of the nuclei will be more stable than others. The unstable isotopes are called *radioisotopes*, as in order to reach a more stable state they emit energy in the form of radiations.

An isotope of sodium, for example, that occurs naturally has 11 protons and 12 neutrons in its nucleus. The shorthand description of this isotope is $^{23}_{11}$Na, and sometimes only ^{23}Na. This isotope is stable.

^{24}Na on the other hand is a radioisotope with 11 protons and 13 neutrons and it is widely used in both biology and medicine. It is produced in nuclear reactors by bombarding stable sodium with neutrons.

$$^{23}_{11}\text{Na} + ^{1}_{0}\text{n} \rightarrow ^{24}_{11}\text{Na} + \gamma\text{-rays}. \tag{13.2}$$

As the neutron is uncharged it can readily enter the ^{23}Na nucleus to produce a different atomic species, $^{24}_{11}$Na. This isotope is unstable and it reaches the stable state by emitting a highly energetic β-particle (1.39 MeV) and two photons of energies 1.37 and 2.75 MeV respectively.

$$^{24}_{11}\text{Na} \rightarrow ^{24}_{12}\text{Mg} + \beta^- + \gamma\text{-rays}. \tag{13.3}$$

^{24}Na is unstable because of an excess of neutrons, and one of these in effect decays to give a proton and an electron, leaving the element Mg in place of Na.

Unstable nuclei are characterized both by the radiations they emit and their decay time, or *half-life*. Since radioactive decay is a random process, the probability that an unstable atom changes its state is constant within a given period of time. This probability is unaffected by the history of the atom, its chemical or physical state, or by the passage of time. The rate at which a quantity of radioisotope decays is therefore proportional to the number of unstable atoms present (N), i.e.

$$\frac{dN}{dt} = -\lambda N \tag{13.4}$$

where λ is the decay rate constant.

If N_0 atoms are present at $t = 0$, the number present at some time t s later will be given by the equation

$$N_t = N_0 \exp(-\lambda t). \tag{13.5}$$

This is in fact the solution to the above differential equation and can be written in the form

$$\log_e \frac{N_t}{N_0} = -\lambda t.$$

The half-life is simply the time taken ($t_{1/2}$) to decay from N_0 to $\frac{1}{2}N_0$ and the relationship between $t_{1/2}$ and λ is given by

$$\log_e \tfrac{1}{2} = -\lambda t_{1/2}$$

$$\text{or } t_{1/2} = \frac{0.693}{\lambda}. \tag{13.6}$$

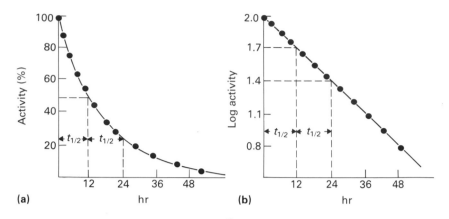

Fig. 13.1. Activity in a sample containing ^{42}K measured as a function of time. (a) linear, (b) semi-logarithmic scale.

Table 13.2. Physical data of radioisotopes.

Isotope	Half-life	Radiation	Energy
^{137}Cs	30 yr	β^-	0.51 and 1.17 MeV
		γ	0.66 MeV
^{45}Ca	165 d	β^-	0.25 MeV
^{14}C	5760 yr	β^-	0.16 MeV
^{3}H (tritium)	12 yr	β^-	18 keV
^{125}I	60 d	β^-	35 keV
^{131}I	8 d	β^-	0.61 MeV
		γ	0.36 MeV
^{15}O	2 min	β^+	0.511 MeV
			γ from β^+
^{32}P	14 d	β^-	1.7 MeV
^{40}K	1.3×10^9 yr	β^-	1.3 MeV
		γ	1.5 MeV
^{42}K	12 hr	β^-	2.0 and 3.6 MeV
^{222}Rn	3.8 d	α^+	548 MeV
			γ from positron
^{22}Na	2.6 yr	β^+	0.51 MeV
		γ	1.28 Mev
^{24}Na	15 hr	β^-	1.39 MeV
		γ	1.37 and 2.75 MeV
^{89}Sr	51 d	β^-	1.46 MeV
^{90}Sr	28 yr	β^-	0.45 MeV
^{35}S	87 d	β^-	0.167 MeV
^{99}T	6 hr	γ	140 keV
^{255}U	7×10^8 yr	α^+	4.2–4.6 MeV
^{133}Xe	503 d	β^-	0.34 MeV
		γ	81 keV

The energies of the radiations are given in electron volts and the SI equivalent is the joule, where 1 eV $= 1.6 \times 10^{-7}$ J. Note that the positron has only a transient existence and 2 γ-photons are emitted in opposite directions when it encounters an electron (annihilation energy).

$t_{1/2}$ or λ can be readily obtained experimentally by measuring how the activity in a radioactivity sample decays with time. For example, the activity in a sample containing the isotope ^{42}K decays quite rapidly with time and a graph of activity (counts s^{-1}) against time has an exponential form (Fig. 13.1). The half-time is close to 12 h. The data can also be plotted on a logarithmic scale, when a straight line is obtained.

The unit of radioactivity for an isotope is basically the number of disintegrations occurring in the material per second and this was chosen as it is proportional to the number of radioactive atoms present. The activity unit first chosen was the *curie* and corresponds to the number of disintegrations per second produced by one gram of radium.

1 curie (Ci) $= 3.7 \times 10^{10}$ disintegrations per second. The new SI unit is the becquerel (Bq) and is equivalent to one disintegration per second, i.e.

$$1 \ \mu\text{Ci} = 3.7 \times 10^4 \ \text{Bq}$$

or 0.1 μCi $= 3.7$ kBq etc.

13.4 Interactions of radiations with matter

Because they are highly energetic, radiations interact very strongly with atoms and molecules in their path. They have in fact sufficient energy to cause *ionization* and so are often referred to as *ionizing radiations*. The end product of this ionization is radiation damage in the material which they traverse. If the material is an animal cell, however, the end result could be mutilation or death for the animal.

X-ray photons are absorbed by materials by transferring their energy to atomic electrons in much the same manner as light photons. In the case of X-rays and γ-rays, however, the electrons gain sufficient energy to leave the atom altogether. The same absorption law holds for both cases

$$I_x = I_0 \exp\left(-\mu x\right) \tag{13.7}$$

where I_0 is the intensity of X-rays at the surface of the material and I_x is the intensity at some distance x from the surface. μ, the absorption coeffieient of the material, increases with the *atomic number*, i.e. number of protons in the nucleus, of the material. The denser the material, the greater is the attenuation of original intensity and, as bones are much denser than the surrounding tissue, they are well defined on X-ray photographs. The depth of penetration of X-rays increases with energy (Table 13.1).

γ-rays have more than one mode of interaction with matter. The rays can ionize an atom in much the same way as X-rays, or they can penetrate the nucleus if they have sufficient energy. When this occurs the γ-ray energy is taken up in producing an electron and its antiparticle, the positron. These particles leave the nucleus and create further ionizations. In this case, the energy of the γ-ray creates two particles and this awesome feat is reversed when a positron and an electron collide — they are annihilated and γ-rays are produced.

The charged particles commonly encountered (α- and β-particles and positrons) interact with the electric fields of the external atomic electrons and again the latter can be given sufficient energy to leave the atom and produce a so-called *ion pair* consisting of an electron and an ionized atom.

Neutrons are very damaging. Because they are uncharged, they can easily penetrate the nucleus, render it unstable, and so produce further radiations.

13.5   Biological effects of radiation

The principal material of which most cells are composed is of course water and much effort has been expended in order to elucidate the effect of radiations on water. It now appears that when the primary ion pairs are produced, reactive ions and free radicals are formed in the water

$$H_2O + \text{absorbed energy} \rightarrow H_2O^+ + e^- \tag{13.8}$$

$$H_2O + e^- \qquad\qquad \rightarrow H_2O^- \text{ (solvated electron)} . \tag{13.9}$$

It is proposed that the highly reactive species so formed then interact with, and indeed damage, other molecules, e.g. DNA, which may not have been themselves hit by the ionizing radiations.

The main unit used to quantify the interaction of radiation with matter is the *rad* (Radiation Absorbed Dose) and this is the quantity of radiation that will result in an energy absorption of 10^{-2} J per kg of tissue. The SI unit for absorbed dose, the gray (Gy) is becoming increasingly used and 1 Gy = 100 rad.

As far as radiation damage is concerned, not only is the total energy absorbed by the tissue important, but also how localized is the energy absorption. It appears that densely ionizing particles are the more damaging and this is probably because in many cases several hits are required on one site in a molecule to damage it permanently. The *linear energy transfer* (LET) of a particle determines the radiation damage, and the LET is the energy deposited per metre of particle path length. It has been found that LET increases as the square of the particle charge and decreases as the square of the velocity, so α-particles are relatively more damaging than β-particles, while low energy βs are more damaging than high energy βs.

The LET determines the *relative biological effectiveness* (RBE) of a radiation type. The RBE of γ- and X-radiation is about equal and taken as 1. The RBE of fast moving β-particles is also 1, while that of very slow-moving β's is 2. The RBE value for slow neutrons and α-particles are 3 and 10 respectively. This means that if 1 rad of γ-rays produces a certain effect in an animal or tissue, then only 0.5 rad of slow βs or 0.1 rad of α-particles will produce the same effect.

However, even with the RBE of the various radiations taken into account, we cannot simply state the biological effect of a dose of, say, 1 rad. This is because some animals are more radiosensitive than others, just as different tissues also vary. The total body dose (X-or γ-rays) that will kill 50% of a certain population of animals (the so-called LD_{50}) is 3000 rad for newts and 250 rad for guinea pigs, whereas the LD_{50} for man, computed from the Nagasaki and Hiroshima data, is about 650 rad. If certain very radiosensitive parts of the body, notably the spleen, are shielded, then the LD_{50} increases.

However, even taking species and tissue variation into account, we still have not solved the problem of assessing the effect of a certain radiation dose, because not only is the total dose important, but so is its administration. Table 13.3 shows the total dose necessary to produce in people a fixed intensity of skin reddening at 2 weeks following X-radiation. Narrow beams were used, so the total irradiated area was very small. The fact that the required total dose increased markedly as the dose rate decreased implies that there are recovery processes in the skin.

Dose rate (rad min^{-1})	Irradiation time (min)	Total dose (rad)
500	1	500
50	15.5	780
5	260	1300
0.5	4500	2250

Table 13.3. X-ray dose to produce skin reddening (from Alexander, 1965).

Skin reddening, erythema, is just one of the many gross effects produced by radiation. Relatively low doses to the whole body, as little as 25 rad (X-ray equivalent), will produce a significant drop in the white cell count of the blood and so, if there is any doubt about possible exposure to radiation, a simple count should be carried out. The red blood cells on the other hand are more radioresistant.

Sterility in men occurs after an exposure to the gonads of about 4000 rad while in women it occurs after 1000 rad. If 400 rad are received, men remain fertile for many weeks, are then sterile for several months, and finally recover fertility. The reason for this is that sperms are resistant but the sperm precursor cells are not and the months of sterility are ended only by the repopulation of the testes with sperm

	Dose to chest (mR)	Dose to gonads (mR)
Hospital chest X-ray	5–30	0.1–1
Mass radiography	100–500	1–5

Table 13.4. Medical radiation doses. The dose ranges are for modern, well-shielded machines.

precursor cells. For anyone who is worried by this, the doses likely to be encountered during an X-ray examination are tabulated (Table 13.4).

Great care has to be taken in the radiological examination of women of childbearing age because during a radiological examination of the pelvic region itself, the dose to the gonads can exceed 1 rad. Experiments with mice, for example, have shown that if 200 rad are received to the pelvic region within a few days after conception, no live births occur as the zygote is very radiosensitive. However, if the embryo is irradiated during the differentiating stages, the birth of live but abnormal offspring will result and as little as 5 rad can produce abnormalities. Equivalent irradiation in the latter stages of pregnancy produces very few abnormalities.

From the above observations it is possible to draw the following conclusions about the relative radiosensitivity of different cells in the body: the most radiosensitive cells are those which (i) have the highest division rate, (ii) retain the capacity for division the longest, (iii) are the least differentiated.

However, not only do radiations induce gross changes in tissues, they also produce *mutations* in the genetic material of the animal and these changes may only be manifest generations later. For radiation death, there is clearly a lower limit below which it is unlikely for any deaths to occur (Fig. 13.2a). However, careful experiments performed on large numbers of mice seem to show that the number of radiation in-

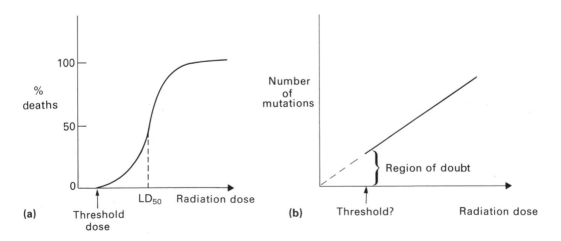

Fig. 13.2. See text for explanation.

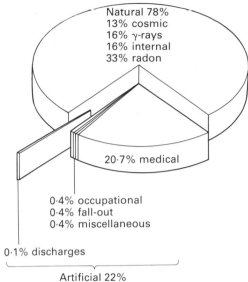

Natural 78%
13% cosmic
16% γ-rays
16% internal
33% radon

20·7% medical

0·4% occupational
0·4% fall-out
0·4% miscellaneous

0·1% discharges

Artificial 22%

Fig. 13.3. Sources of radiation to the UK population (pre-Chernobyl). The average dose is approximately 100 mrad y^{-1} (1000 μGy y^{-1}) and 78% of that originates from natural sources such as radon gas (e.g. in granite buildings) and the isotope ^{40}K which is present in cells and tissues. The dose from atmospheric fallout had been declining steadily until Chernobyl.

duced mutations (Fig. 13.2b) increases linearly with dose and so genetically any exposure to radiation can be considered harmful.

As X-ray photographs have to be taken, and we often wish to work with radioactive materials, we have to decide on certain maximum permissible levels. In considering genetic effects, we are not concerned with individuals, but with the average dose to the whole population. The natural background level of 0.1 rad per year is the normal exposure to the population in Britain for example (Fig. 13.3), and in fact there appears to be no detectable increase in mutations in those regions e.g. Kerala in India, where the background rate is at least twice normal. Thus it seems that a population can tolerate a further 0.1 rad per year or, over a generation time of 30 years, 3 rad per generation. A few special workers are permitted a greater exposure, but they contribute so little to the gene pool that the sum effect may be neglected. The main threat to our genetic heritage at present comes not from isotope users, atomic energy authorities, or even bomb tests, but from the therapeutic and diagnostic uses of X-rays. Fortunately this has been recognized for a long time and, in the UK at least, the total dose to the population from medical X-rays adds only 20% to the natural background.

Problem 13.1

20×10^6 Ci of activity were released into the atmosphere at Chernobyl (compared with 30×10^3 Ci released during the Windscale reactor fire in 1957). Show that this radiation uniformly distributed over the surface of a sphere of radius 6.3×10^3 km gives an activity of approximately 1.5×10^3 Bq m^{-2}.

13.6   Radiobiological consequences of nuclear fission

Nuclear fission occurs when a large atomic nucleus such as ^{235}U or ^{239}Pu is bombarded with neutrons. The unstable nucleus produced disintegrates to give two nuclei of comparable masses and large quantities of energy (Q) are released. More importantly, however, neutrons are released and each of these is capable of initiating a further fission so that a chain reaction can be set up.

$$^{1}_{0}n + {}^{235}U \rightarrow {}^{236}U \text{ (neutron capture)}$$

$$^{236}U \rightarrow {}^{141}Ba + {}^{92}Kr + 3{}^{1}_{0}n + Q \text{ (fission of unstable } {}^{236}U).$$

In this case the *daughter* products are barium and krypton, but ^{90}Sr, ^{137}Cs and ^{131}I are also possible daughter nuclei.

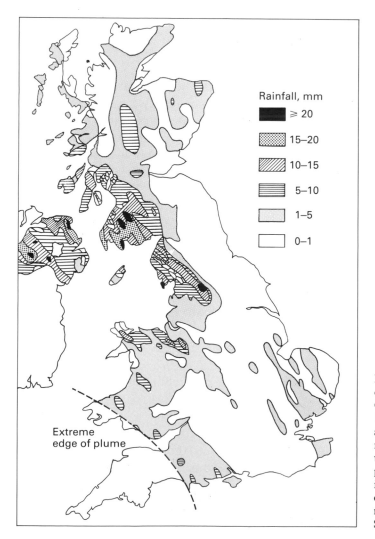

Rainfall, mm

■ ≥ 20

15–20

10–15

5–10

1–5

0–1

Extreme edge of plume

Fig. 13.4. Rainfall in the UK during the passage of the Chernobyl plume, 2–4 May 1986, compiled from radar data and measurements at 4000 rain gauge stations. This is not the total rainfall for that period, only that which intercepted the plume, based on air concentration measurements (from Clark & Smith, 1988).

In nuclear bombs, the fission reaction is uncontrolled, while in nuclear reactors it is controlled mainly by inserting materials into the reactor core that absorb neutrons but do not produce further neutrons, e.g. boron. The reactor vessel is also cooled and the heat energy Q is converted into electrical energy. In the Chernobyl accident, the control processes were over-ridden and vast amounts of radioactivity ($\approx 10^{19}$ Bq) were released into the atmosphere. In many respects the results were more devastating to certain sections of the world community than if the same amount of radiation had been released by a small nuclear bomb.

In most cases nuclear bombs have been detonated underground or in isolated environments. In the latter case the mushroom cloud takes most of the radiation into the upper atmosphere from where it is released back to earth over a very wide area and over a prolonged period of time (1–6 years). In the case of Chernobyl the upward convective forces were smaller and the radioactive air mass was carried north by the prevailing winds until it reached northern Sweden. It then travelled

Table 13.5. Regional averages for deposition of ^{131}I and ^{137}Cs on grass in UK during 3–6 May 1986.

Region	Deposition on grass (Bq m^{-2})		^{131}I/^{137}Cs ratio
	^{131}I	^{137}Cs	
Dry areas			
Berkshire	242	20	12
Channel Islands	151	4	38
Cleveland	70	17	4
Oxfordshire	459	22	21
Light rainfall			
Cheshire	325	93	3.5
Dorset	64	25	2.6
Essex	1345	285	4.7
Gloucestershire	1356	351	2.9
Kent	2346	—	—
Lancashire	1327	358	3.7
Wet areas			
Cumbria	2246	1334	1.7
Dumfries	610	598	1.0
Gwynedd	1009	1235	0.8
Highlands	3604	1164	3.1
Orkney	3159	682	4.6
Strathclyde	3445	2285	1.5
West Yorkshire	1259	680	1.9
North Yorkshire	801	470	1.7

The data were obtained from grass samples since in the UK these values relate to subsequent levels in cow's milk and other foodstuffs. Note that the ^{131}I/^{137}Cs ratio varies from above 20 in dry areas to below 1 in very wet regions (data from Clark & Smith, 1988).

in a southerly airstream to deposit significant amounts of radioactivity over the UK, Germany and Italy. The radioactivity deposited was not uniform. For example, ^{137}Cs, which is deposited mainly in water droplets, fell in the UK over areas with a high rainfall, in Scotland, Cumbria and North Wales. Drier areas along the eastern coast and in the South, on the other hand, received much lower levels. However, dry areas received significant deposits of ^{131}I (Fig. 13.4 and Table 13.5).

The calculation in Problem 13.1 indicates that if the fall-out from Chernobyl had been uniformly distributed over the surface of the earth, the expected average activity would have been 1.5×10^3 Bq m^{-2}. The estimated deposition of ^{137}Cs alone in fact varied from 0.1 in dry areas to $>20 \times 10^3$ Bq m^{-2} in wet areas (Clark & Smith, 1988). There was much less variation in the deposition of ^{131}I, which exists mainly in the vapour form and so does not require rainfall to bring it to earth (Table 13.5). In many respects, therefore, parts of the UK and other European countries received more than their 'fair share' of the fall-out from Chernobyl.

The deposition of large amounts of radioactivity over Lapland was particularly serious. Even before Chernobyl it had been pointed out that Lapland reindeer meat contained almost 100 times more radioactive caesium than that from animals in temperate zones (Thornburn, 1972). It is worthwhile considering in general terms how people are at risk from airborne contamination (Fig. 13.5) and in particular why the Laplanders were so much more at risk.

The major contaminating fission products are ^{90}Sr, ^{137}Cs and ^{131}I. Strontium is chemically similar to calcium and caesium is chemically similar to potassium. The possible routes for entry of the contaminants into our diet are represented diagrammatically in Fig. 13.5 and the critical paths for the three major radioisotopes can be traced.

Iodine is a relatively short-lived isotope (8 days) and direct foliar contamination of terrestrial plants is the major critical path. It is readily absorbed through the gut and although most is taken up by the thyroid gland, a significant fraction enters the milk of lactating animals. This is the major route to people. For example, in the UK where the best data are available (Clark & Smith, 1988), the activity of ^{131}I, assayed in milk a few days after Chernobyl, exceeded 150 Bq l^{-1} in certain wet areas. ^{89}Sr and ^{90}Sr are both produced during nuclear fission but their relative importance varies considerably with time because of their different half-lives (50 days and 27 years respectively). Little or no strontium enters plants through the leaves and it is relatively immobile within the plant. Strontium is mainly taken up through the root system and in the gut its absorption is relatively inefficient compared with that of calcium. As dietary calcium increases, for example, more strontium is excreted. With mature animals, calcium turnover in the body is low, but

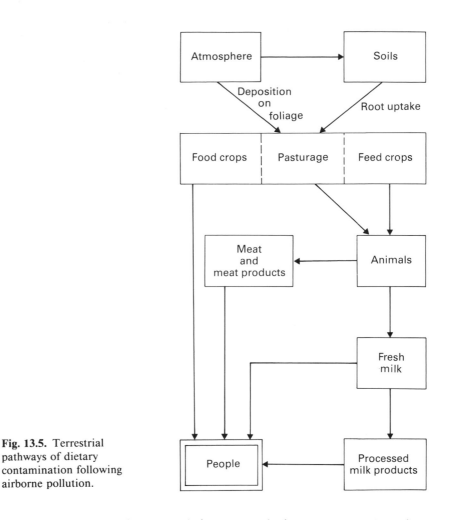

Fig. 13.5. Terrestrial pathways of dietary contamination following airborne pollution.

calcium metabolism is extremely important during pregnancy, lactation and growth. Infants and children are therefore most at risk from strontium contamination.

^{137}Cs is much more of a general risk. It has a relatively long half-life (30 years), can be absorbed into plants directly by the foliar route (especially in lichens) and is highly mobile within the plant. It is concentrated and stored in plant tissues and is rapidly taken up by the gut where it distributes throughout the whole body. In milk samples taken in the UK between 5 and 8 May 1986, the level of ^{137}Cs ranged from over 150 Bq l^{-1} in wet western areas to below 5 Bq l^{-1} in dry eastern regions.

The permitted radiation level for meat in Europe was set at approximately 500 Bq kg^{-1} immediately after Chernobyl and many thousands of lambs from the wet areas of the UK were taken off the market in the succeeding months. One year later the levels in North Wales were generally below 500 Bq kg^{-1}. However, in Lapland, many reindeer were found to have a body burden of 30×10^3 Bq kg^{-1} 12 months after the accident. The Laplanders suffered greatly (a) because of the direction of the prevailing wind, (b) because lichens, the major source of reindeer fodder, efficiently absorb ^{137}Cs through exposed surfaces and from the peaty Arctic soil and (c) because reindeer meat is a staple diet.

13.7 Radiation detectors

Now that we know some of the possible dangers involved in handling isotopes and X-ray sources (Section 13.5) we are better equipped to plan experiments involving isotopes. The actual steps that should be carried out before and during the experiment to ensure that minimum risk is involved to oneself and one's colleagues are described in great detail by Thornburn (1972) and as they involve more common sense than science they will not be repeated here. However, an important part of the planning of an experiment will be described, and that is the background knowledge required to choose the most efficient way of assaying the isotopes (and X-ray sources) used in the experiment. All of the various techniques will record background radiation (Fig. 13.3) even when there is no radioactive sample in the machine and so this background has to be subtracted from the total counts with the sample in place.

Ionization chamber and Geiger–Muller tube

One of the earliest forms of counting devices was the ionization chamber, which is still widely used in hospitals to monitor the radiation from X-ray machines. The chamber (Fig. 13.6) is normally filled with air and has two fixed parallel electrodes across which can be applied a potential difference of several hundred volts. Normally the air inside the chamber does not conduct electricity and so no current will flow through the system. However, when radiations pass through the chamber they will ionize the enclosed gas; the electrons will move to the

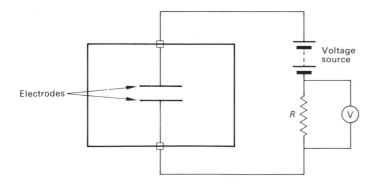

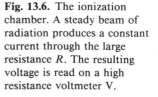

Fig. 13.6. The ionization chamber. A steady beam of radiation produces a constant current through the large resistance R. The resulting voltage is read on a high resistance voltmeter V.

anode, and the positively charged ions to the cathode. If a steady beam of radiation passes through the chamber, a steady current will flow through the resistance R and its magnitude can be calculated from the reading on the high impedance voltmeter V. When X-rays are the radiation source, the dose rate D in röntgens per hour (see problem 13.2) is directly related to the current I by the equation

$$I = 9.25 \times 10^{-14} \times VD \qquad (13.10)$$

where V is the volume of the chamber in cm^3. If the voltmeter has a suitably short time-constant the chamber can be used as a proportional counter to count single particles as each of these will produce a single voltage pulse.

The sensitivity of such an ionization chamber can be increased by raising the potential difference between the plates (HT in Fig. 13.7) to about 1 kV and by filling the chamber with special gases. The modifications were introduced by Geiger and Muller in 1920 and in modern versions of the G–M tube the gases are normally argon and ethyl alcohol. A thin central wire acts as an anode while the metal of the chamber (earthed) is the cathode. The radiations enter through a thin metal or mica window at one end so that absorption before entering is reduced (Fig. 13.7). When an ion pair is produced the electron moves with a high velocity towards the central wire and the positive argon ion moves slowly to the cathode. The fast moving electron ionizes further argon atoms and an electronic avalanche is set up, the size of which is independent of the energy and position of the initial ionization. This results in an anode current. As the electrons from the avalanche are collected, the slowly moving argon ions act as an electrostatic screen and this reduces the field at the wire to a value below that necessary for ionization by collision, so that the discharge should cease. However, there is the possibility that the positive argon ions can eject electrons

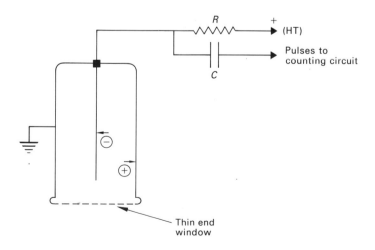

Fig. 13.7. The Geiger–Muller tube. As the G–M tube is normally used to count single pulses the time constant of the circuit RC must be short.

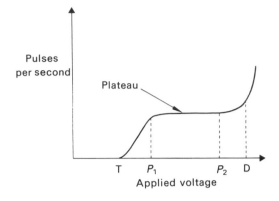

Fig. 13.8. Operating characteristics of the G–M tube. The tube starts to count at T, the threshold voltage. The operating region (or G–M plateau) is between the voltage at P_1 and P_2. At higher voltages, D, continuous discharge occurs (and the tube may be ruined).

from the cathode and to prevent this the alcohol is added to quench the discharge, which it does because of its low ionization potential. The argon ions on their journey to the cathode are neutralized by acquiring an electron from the alcohol molecules. The ions reaching the cathode are then alcohol ions and although they themselves are neutralized at the cathode, the energy is absorbed in dissociating the alcohol molecule rather than in producing further electrons from the cathode. The discharge thus ceases when the field around the anode has fallen sufficiently.

The advantages of the G–M tube lie in its sensitivity and in its operating characteristics (Fig. 13.8). The count rate is constant over a relatively large range of applied voltages. Although the G–M tube can be used to count relatively low energy particles, e.g. β from $^{14}_{6}C$, it is useless for the βs from tritium ($^{3}_{1}H$) as these have insufficient energy to penetrate even the thin mica window. The G–M tube is also inefficient at counting γ-rays and for these a different counting technique is employed.

A serious disadvantage of the G–M tube is that it has a relatively long insensitive time following each current pulse. This is made up of a dead time during which the electric field near the anode has dropped below the counting threshold and a recovery time during which pulses of reduced size are produced. The insensitive time, approximately 300 μs, is usually increased and set at a definite value by the addition of an external electrical circuit. At high counting rates a correction for lost counts must be applied.

Problem 13.2

(a)   Given that 1 röntgen is defined as that dose of radiation which produces 2.1×10^9 ion pairs in 1 cm^3 of air, derive the relationship between current I and dose rate D given in equation (13.10).

1 electronic charge $= 1.6 \times 10^{-19}$ C.

(*b*)   Thc rad is the radiation absorbed dose and is defined as an energy absorption of 10^{-2} J kg^{-1} material. What is the dose absorbed by air exposed to 1 röntgen of X-rays? (34 eV is associated with 1 ion pair; density of air = 1.29×10^{-3} g cm^{-3}.)

Scintillation counters

One of the earliest observations of nuclear physics was that certain materials emit absorbed radiation in the form of tiny light pulses or scintillations. The first scintillation counter consisted of a zinc sulphide screen viewed by a low power microscope. It was found that for accurate measurements of activity with this system, the scintillation counts had to be corrected for a dead time, which in this case was the blink time of the observer. The design of a modern scintillation counter is shown in Fig. 13.9.

In one form of scintillation counter a crystal, usually composed of sodium iodide and activated by thallium, is used. The radiation produces light pulses along its track and the pulses are counted using a

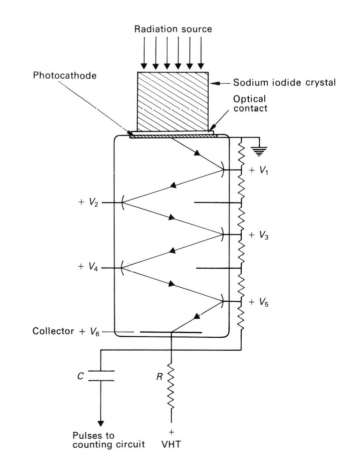

Fig. 13.9. Crystal scintillation counter. The crystal and photomultiplier are built into a light-proof box and the photomultiplier is surrounded by a metal screen to prevent interference from magnetic fields.

photomultiplier tube. The impinging light photons drive electrons from the surface of a photocathode at the first stage of a photomultiplier. The initial pulse of electrons is amplified in the photomultiplier which consists of a series of electrodes. Application of an increasing positive potential between successive electrodes accelerates the electrons from one electrode to the next and at each impact 2–5 secondary electrons per primary may be produced. The final highly amplified current is collected at the anode and on passing through a suitable resistor can produce a pulse of several volts.

These counters have certain advantages over the G–M tube as they have a very short insensitive time and they also count γ-rays more efficiently, because of the greater density of the sensitive materials. The photomultiplier does, however, require a higher and much more stable operating voltage. In another form the radioactive sample is mixed with a liquid scintillator material, which has two main components: the solvent, usually xylene and the primary solute, often diphenyloxazole. Most of the energy is absorbed from the radiation by the solvent and then transferred to the solute, which actually emits the light. Often a small amount of secondary solute is added to shift the wavelength of the emitted photons in order to increase the transparency of the scintillator to the light photons. As the radioactive compounds are dissolved in the scintillator the counting efficiency, especially for particles of low energy (β-particles from ^{3}H and ^{14}C) is relatively high. The sample is usually placed over a photomultiplier.

As the magnitude of the light flashes, and hence of the final output, depends on the energy of a particle, the final electronics can be arranged so that only pulses of a certain height are counted. With several electronic discriminators or gates set at different levels, a corresponding number of different isotopes can be counted in the same bottle, provided their radiations have sufficiently different energies, e.g. ^{3}H, ^{14}C, ^{32}P.

Cerenkov counting

Machines that assay by liquid scintillation methods can also be used to count radioactive samples in aqueous solution without any scintillator, provided the particles emitted are charged and are moving with sufficiently high velocities. When the velocity of the particle is greater than the velocity of light in the medium in which the particle is travelling, light photons are produced as a result of an electromagnetic shock wave induced in the medium in which the particle is travelling (analogous to a supersonic bang). The isotopes ^{22}Na, ^{24}Na and ^{42}K, widely used in biology, can be assayed by this technique as they emit β-particles of a sufficiently high energy.

Autoradiography

To satisfy the needs of nuclear physicists, photographic emulsions have been developed that have a small grain size and low background on developing. Such films are now being widely used in biology to locate small numbers of specific radioactively-labelled atoms or molecules in tissues. This technique gives the highest resolution with particles of low energy (β-particles from ^{3}H and ^{14}C) where the length of track of silver atoms comprising the latent image is short. On developing, the latent images are amplified and appear as black grains of metallic silver.

In this technique, a thin slice of tissue that has previously been

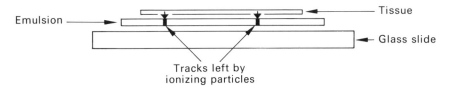

Fig. 13.10. Autoradiography. When the emulsion is developed, the tracks are black against a lighter background.

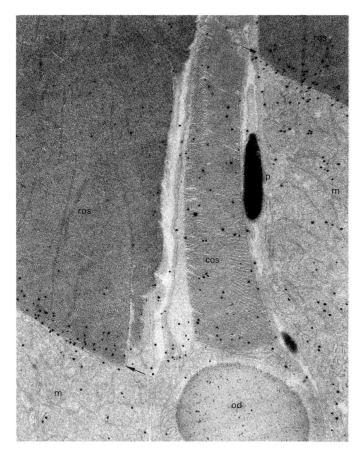

Fig. 13.11. One day after injection. The diffuse distribution of renewal protein in the principal cone outer segment (cos) contrast markedly with the discrete and heavy labelling of a small group of discs near the base of the rod outer segment (ros). The tracks appear black as only silver grains remain on the emulsion after the developing process. Pigment granules (p) surrounding the cells, mitochondria (m) in the inner segments of the rods, and oil droplets (od) at the base of the cones can also be clearly seen. (Electron micrograph, autoradiogram ×14 000.) (Micrograph kindly supplied by Dr R.W. Young, University of California at Los Angeles, with permission.)

treated with an appropriate radioactive substance is placed over the film (Fig. 13.10) and left for a suitable, often very prolonged, time. After development, the film can be examined under the optical or electron microscope to determine the distribution of radioactive substance in the tissue.

Young (1969) has used autoradiography to show that there is a continual turnover of discs in the outer segments (rods) of vertebrate photoreceptor cells (Fig. 13.11). After a short exposure (pulse labelling) to a mixture of amino acids injected into the blood, it can be seen by sacrificing animals at different times that the amino acids are incorporated into the disc membranes (separate experiments show that it is incorporated into the protein part of the photopigment rhodopsin) starting at the discs near the inner segments. These labelled discs are gradually pushed upwards by non-labelled discs until they reach the pigment epithelium where they are broken down. Cones, on the other hand, are diffusely labelled from the beginning, suggesting that all regions of the membrane are equally accessible to the source of label in the inner segment.

13.8   Medical imaging using radioisotopes

Over the past 40 years radioistopes have been widely used in medicine both as tracer substances (e.g. ^{24}Na) to monitor sodium movements in the kidney, for example, and as radiotherapeutic agents in the treatment of cancer. In the last five years, isotopes have been increasingly used to form images of tissues and organs of the body. At present, the two main techniques are *positron emission tomography (PET)* and *single photon emission computed tomography (SPECT)*.

Positron emission tomography (PET)

When an electron and its antiparticle, the positron, meet, matter is converted into energy in the form of two γ photons which are emitted in opposite directions. The photons emitted in the region of interest are captured by a bank of scintillation counters embedded in an annulus with the region at its centre (Fig. 13.12). The paths of the photons can be traced by computer and a two-dimensional image of the source of radiation can be built up. An exciting aspect of this type of imaging is that it can be used to give dynamic information about the flow of blood in the brain or heart, for example. This technique has been used to great effect to map out the changes in blood flow that occur in certain regions of the visual cortex when the (human) subject eye is stimulated with different patterns (Figs 13.13 and 13.14). In this case normal volunteers received a single injection of water labelled with the positron-emitting

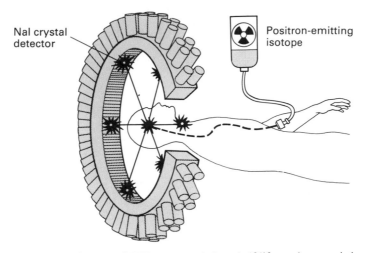

Nal crystal
detector

Positron-emitting
isotope

Fig. 13.12. Diagram of PET scanner. A short half-life, positron-emitting isotope is injected into the patient. In this example brain function is being investigated so the head is placed at the centre of the annulus of detectors. Some commercial models have over 50 NaI detectors in the annulus. Since 2 γ-photons are emitted in opposite directions when a positron and electron collide, their ray paths can be traced back to the emission site (after Sochurek, 1987).

Fig. 13.13. Test stimuli presented to the volunteers in a brain-function experiment. The test consisted of checkerboard rings around a central fixation point. Three stimulus conditions were chosen and the one shown was the most peripheral stimulus and extended from 5.5° to 15.5° around the visual axis. The innermost stimulus consisted of a checkerboard annulus extending from 0.1° to 1.5° (macular) and the intermediate stimulus extended from 1.5° to 5.5°. These stimuli were chosen as they produce very strong activity in the retina and visual cortex. The control stimulus consisted solely of a fixation point. The changes in brain blood flow produced by the three stimulating conditions are shown in Fig. 13.14b–d (from Fox *et al.*, 1986).

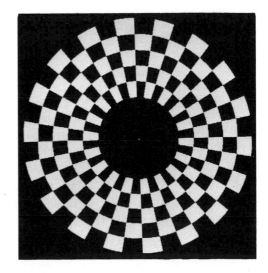

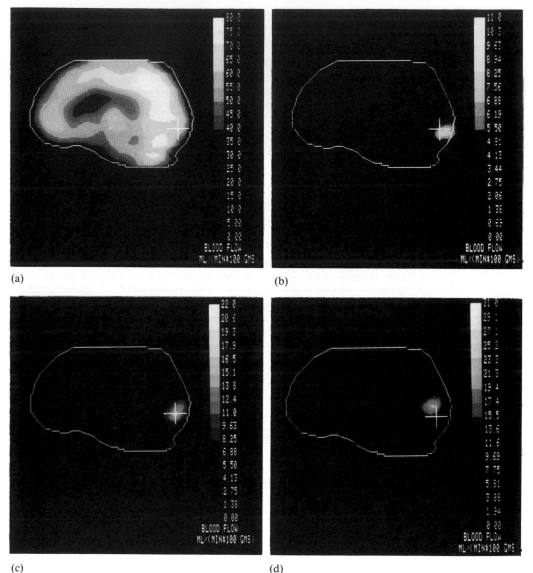

(a)

(b)

(c)

(d)

Fig. 13.14. PET images obtained from midsagittal sections of the brain followed injection of water labelled with ^{15}O into a peripheral vein. The units of blood flow are ml g^{-1} min^{-1} and the anterior brain is to the left. All images were formed by summing the images of six subjects within a standardized stereotactic coordinate space (see Fox *et al.* 1986 for further details). The cross marks the intersection of anterior and posterior brain commisures. (a) Resting state image of cerebral blood flow. This image was used to compute the brain boundary in (a–d). (b) Macular response focus in a subtraction image format and is the most posterior of the responses.
(c) Perimacular response focus, intermediate in location between macular and peripheral responses. (d) Peripheral response focus. (Images kindly provided by Dr P.T. Fox and used with permission.)

^{15}O isotope. The isotope circulated into the brain within 10–20 s and then a 40 s PET scan recorded the distribution of ^{15}O within the brain from which images of regional cerebral blood flow were calculated. Figure 13.14(a) represents the summation of the images of all six subjects participating in the test. Images were then formed (by subsequent injections) of the subjects participating in three defined visual tasks (Fig. 13.13). When the control images were digitally subtracted from the test images it was apparent that stimulating different regions of the retina, and hence of the visual cortex, leads to different patterns of blood flow within the brain.

The importance of this work was not only that it represents one of the first examples of functional brain mapping by PET but, more importantly, it demonstrated that the resolution of the PET technique can be greatly improved by the image subtraction method. Recently this method has been extended to include studies of those areas of the brain involved in processing single words.

As most of the radioisotopes used are extremely short-lived, the PET scanner has to be housed near a device for producing isotopes, such as a cyclotron source or a nuclear reactor.

Single photon emission computed tomography (SPECT)

This technique, like PET, provides tomographic images after administration of a radioisotope. In this case, however, a single photon emitting species is used such as ^{133}Xe, ^{99}Tc or ^{123}I and the technique is much less expensive than PET. The detection device is much simpler but there is less spatial and temporal resolution and there are fewer radiopharmaceuticals available. However, this technique gives acceptable images of the brain and is being used, for example, in the study of Alzheimer's disease (Anderson *et al.*, 1986).

References

Alexander, P. (1965) *Atomic Radiation and Life*. Pelican, Middlesex.

Andersen, A.R., Paulson, O.B. *et al.* (1986) Tomographic brain imaging of cerebal blood flow in dementia: a pilot study. *Modern Trends in Aging Research, Colloque.* INSERM-EURAGE, **147**, 469–72.

Clark, M.J. & Smith, F.B. (1988) Wet and dry deposition of Chernobyl releases. *Nature*, **332**, 245–8.

Fox, P.T., Mintun, M.A., Raichle, M.E., Miezin, F.M., Allman, J.M. & van Essen, D.C. (1986) Mapping human visual cortex with positron emission tomography. *Nature*, **323**, 806–9.

Sochurek, H. (1987) Medicine's new vision. *National Geographic*, **171**, 2–14.

Thornburn, C.C. (1972) *Isotopes and Radiation in Biology*. Butterworths, London.

Young, R.W. (1969) A difference between rods and cones in the renewal of outer segment protein. *Investigative Opthalmology*, **8**, 222–31.

Further reading

Brown, B.H. and Smallwood, R.H. (1981) *Medical Physics and Physiological Measurement.* Blackwell Scientific Publications, Oxford.

Coggle, J.E. (1971) *Biological Effects of Radiation.* Wykeham, London.

Dendy, P.P. & Heaton, B. (1987) *Physics for Radiologists.* Blackwell Scientific Publications, Oxford.

Petersen, S.E., Fox, P.T., Posner, M.I., Mintun, M. & Raichle, M.E. (1988) Positron
emission tomographic studies of the cortical anatomy of single-word processing.
Nature, **331,** 585–9.

Reivich, M. & Alavi, A. (1985) *Positron Emission Tomography.* A.R. Liss, New York.

Appendices

1 The International System of Units (SI)

Table A1. Defined SI units.

Physical quantity	Name of unit	Symbol for unit
Mass	kilogram	kg
Length	metre	m
Time	second	s
Electric current	ampère	A
Thermodynamic Temperature	kelvin	K

Table A2. Special names and symbols for SI derived units.

Physical quantity	Name of unit	Symbol for unit	Definition of unit	Equivalent form
Energy	joule	J	$kg\,m^2\,s^{-2}$	Nm
Force	newton	N	$kg\,m\,s^{-2}$	$J\,m^{-1}$
Pressure	pascal	Pa	$kg\,m^{-1}\,s^{-2}$	$N\,m^{-2}$
Power	watt	W	$kg\,m^2\,s^{-3}$	$J\,s^{-1}$
Electric charge	coulomb	C	As	As
Electric potential difference	volt	V	$kg\,m^2\,s^{-3}\,A^{-1}$	$J\,A^{-1}\,s^{-1}, J\,C^{-1}$
Electric resistance	ohm	Ω	$kg\,m^2\,s^{-3}\,A^{-2}$	$V\,A^{-1}$
Electric capacitance	farad	F	$A^2\,s^4\,kg^{-1}\,m^{-2}$	$A\,s\,V^{-1}, C\,V^{-1}$
Magnetic flux	weber	Wb	$kg\,m^2\,s^{-2}\,A^{-1}$	$V\,s$

Table A3. Units to be allowed in conjunction with SI.

Physical quantity	Name of unit	Symbol for unit	Definition of unit
Volume	litre	1	$10^{-3}\,m^3 = dm^3$
Radioactivity	curie	Ci	$37 \times 10^9\,s^{-1}$
Energy	electron volt	eV	$1.6 \times 10^{-19}\,J$

Table A4. Examples of units contrary to SI with their equivalent.

Physical quantity	Name of unit	Equivalent form
Length	ångström	$10^{-10}\,m$
Force	dyne	$10^{-5}\,N$
Pressure	atmosphere	$10^5\,N\,m^{-2}$
	torr	$133\,N\,m^{-2}$
Energy	erg	$10^{-7}\,J$
	calorie	$4.2\,J$

Table A5. Prefixes for SI units. These prefixes may be used to construct decimal fractions or multiples of units.

Fraction multiple	Prefix	Symbol	Multiple	Prefix	Symbol
10^{-1}	deci	d	10	deca	da
10^{-2}	centi	c	10^2	hecto	h
10^{-3}	milli	m	10^3	kilo	k
10^{-6}	micro	μ	10^6	mega	M
10^{-9}	nano	n	10^9	giga	G
10^{-12}	pico	p	10^{12}	tera	T

Table A6. SI supplementary units.

Physical quantity	Name of unit	Symbol for unit
Plane angle	radian	rad
Solid angle	steradian	sr

Further reading

Quantities Units and Symbols (1971) Royal Society, 6 Carlton House Terrace, London.
van Assendelft, O.W., Mook, G.A. & Zijlstra, W.G. (1973) International system of units (SI) in physiology. *Pflügers Arch*, **339**, 265–72.

2   Vector algebra

Vector addition has been dealt with in Chapter 2; here we shall deal with *vector multiplication*.

1   The *scalar product* of two vectors **A** and **B** is defined by the relationship

$$\mathbf{A}.\mathbf{B} = AB \cos \theta$$

where θ is the smaller of the two possible angles between the vectors (Fig. A1).

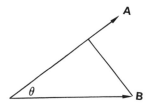

Fig. A1. The projection of **B** on **A** has magnitude $B \cos \theta$ and this multiplied by the magnitude of **A** gives the scalar product $AB \cos \theta$.

With this definition of scalar product, a number of important physical quantities can be described as the scalar product of two vectors, e.g. mechanical work (Chapter 2) and electrical potential (Chapter 10).

2   The *vector product* of two vectors **A** and **B** is written as **A** $\times$ **B** and is another vector **C** where the magnitude of **C** is given by

$$C = AB \sin \theta$$

where θ is the angle between **A** and **B** (Fig. A2). The direction of **C** is perpendicular to the plane of the two vectors **A** and **B** and is defined as follows:

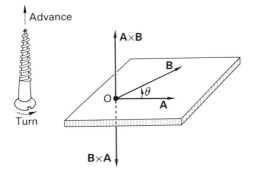

Fig. A2. Vector products of **A** and **B**. The product **A** $\times$ **B** points upward, while the product of **B** $\times$ **A** points downward. The magnitude of both products is $AB \sin \theta$, but **B** $\times$ **A** $= -$**A** $\times$ **B**.

Imagine a right-hand screw with its axes perpendicular to the plane of **A** and **B** and let the screw be turned in the same way the first vector **A** turns when it is rotated through the smallest angle θ that will bring it parallel with the second vector **B**. The product **A** $\times$ **B** then points in the direction of advance of the screw. The vector product **B** $\times$ **A** has the same magnitude as **A** $\times$ **B** but points in the opposite direction.

An example of a quantity that is a vector product is the force on a moving charge in a magnetic field (Chapter 12).

Answers to Selected Problems

Problem 2.2

(1) Kinetic Energy $= \frac{1}{2} m v^2$
$$= 8.8 \times 10^2 \text{ J}$$
Power $\qquad = 8.8 \times 10^2 \text{ J s}^{-1}$
$$= 0.88 \text{ kW}$$
(2) Deceleration $\quad = v/t$
$$= 5 \text{ ms}^{-2}$$
$$\approx 0.5 \text{ g}$$

Problem 2.3

(1) Apply equation 2.9b, with $\quad v_t = 0$
$$\text{i.e. } v_o^2 = 2as$$
$$\therefore a = 14.3 \text{ ms}^{-2}$$
Now, deceleration due to gravity $\quad = 9.8 \text{ ms}^{-2}$
$\therefore$ deceleration due to air resistance $= 4.5 \text{ ms}^{-2}$
Which is approximately half the gravitational deceleration
(2) Acceleration $\quad = v/t$
$$= 1000 \text{ ms}^{-2}$$
which is approximately 100 g
(3) Kinetic energy $= 0.23 \times 10^{-6} \text{ J}$
(4) Flight is powered over a period of 10^{-3} s, hence power input required is 0.23 $\times 10^{-3}$ W.
Maximum output from muscles $= 5.4 \times 10^{-6}$ W, which is chiefly insufficient to power the jump.
(5) Answer in text.

Problem 3.4

(a) This is known as the 'greenhouse effect' and it arises because the illuminated contents of the pack will reradiate longer wavelength radiation than that absorbed (equation 3.9). Because plastics (and glass) absorb these long wavelength radiations, the contents will ultimately heat up.
(b) A metal foil pack with a description or picture of contents would be all that is required, but this would be unpopular with sceptical shoppers. An alternative pack could have a transparent window on the bottom face, with an instruction on the pack to place that face downwards.

Problem 4.1

(a) $1.3 \times 10^4 \text{ Nm}^{-2}$
(b) i. $0.8 \times 10^4 \text{ Nm}^{-2}$
 ii. $1.8 \times 10^4 \text{ Nm}^{-2}$

Problem 5.1

Apply equation 5.1 to the data to give the answer 'yes'.

Problem 5.2

(1) Linear velocity of flow in lumen $= 2.3 \times 10^{-4}\,\text{ms}^{-1}$
Linear velocity of flow in pores $= 4.6 \times 10^{-4}\,\text{ms}^{-1}$
Pressure drop across a $195\,\mu\text{m}$ length of lumen$= 7.1 \times 10^{2}\,\text{Nm}^{-2}$
Pressure drop across sieve plate $= 9.2 \times 10^{2}\,\text{Nm}^{-2}$
Total pressure drop across $200\,\mu\text{m}$ length $= 16.3 \times 10^{2}\,\text{Nm}^{-2}$
Total pressure drop across 1 m $= 7.85 \times 10^{6}\,\text{Nm}^{-2}$
(2) The total possible osmotic driving force would be given by

$$\Delta\pi = RT\Delta C_{\text{s}}$$

providing the phloem membranes at A were impermeable to sucrose

i.e. $\Delta\pi = 7.5 \times 10^{5}\,\text{Nm}^{-2}$ (assuming $T = 293\,\text{K}$)

The osmotic driving force is clearly insufficient to drive the sucrose solution through the phloem at the given rates.
(3) Almost every possible driving force has been suggested at some time and at present the only non-controversial statement in this troubled field is that there is as yet no consensus of opinion.

Problem 5.3

(a) Linear velocity of flow = volume flow/area
$$= 28 \times 10^{-2}\,\text{ms}^{-1}$$
Kinetic energy density is given by $\frac{1}{2}\rho v^2$
$$= 45\,\text{Nm}^{-2},$$
which is small compared with the arterial pressure of $1.3 \times 10^{4}\,\text{Nm}^{-2}$.
(b) The pressure associated with a linear velocity of flow of $84 \times 10^{-2}\,\text{ms}^{-1}$ is $360\,\text{Nm}^{-2}$ which is certainly sufficient to explain the observed pressure difference.
(c) Kinetic energy created $= 10^{3}\,\text{Nm}^{-2}$.

Problem 5.4

From the continuity equation, the velocity of flow increases by a factor of 5 when the lumen narrows to under one-fifth. Therefore the kinetic energy density of the fluid goes up from $40\,\text{Nm}^{-2}$ to $10^{3}\,\text{Nm}^{-2}$ and the arterial pressure is reduced by this amount.

Problem 5.5

From equation 5.16, $N_R = 1500$, which is less than the critical number. During heavy exercise, when the velocity increases fivefold, then $N_R = 7500$, which is greater than the critical number.

Problem 6.1

From equation 6.6, $h = 0.7$ m (maximum)
Other possible mechanisms are root pressure (osmotic driving force due to a gradient of solute concentration across root cell membranes), and transpiration.

Problem 6.2

Upward surface tension force $= 42 \times 10^{-5}$ N
Downward gravitional force $= 24.5 \times 10^{-5}$ N
When the surface tension is reduced by detergents,
upward force $\qquad = 24 \times 10^{-5}$ N

Problem 7.1

(a) Answer given in text
(b) 10%

Problem 7.2

(a) The energy (E') falling on an object a distance r from the source is inversely proportional to r^2

i.e. $E' = \dfrac{kE}{r^2}$

Suppose a fraction of this energy is reflected from the object

i.e. $E'' = \dfrac{kfE}{r^2}$

Again from the inverse square law, the proportion of this energy falling on the source will be inversely proportional to r^2

i.e. $E''' = \dfrac{kE''}{r^2}$

(assuming the bat moves a negligible distance during the process). Hence the intensity of the echo is inversely proportional to r^4.
(b) Ratio of echo intensities from small object $= (r_1 / r_2)^4$
where $r_1 = 0.3$ m and $r_2 = 0.25$ m
Hence $I_2/I_1 = 2.06$
Similarly, ratio of echos from large object $= 1.2$

Problem 7.3

(a) Suppose that v is the relative velocity of the bat and object. Distance between bat and object at first echo $= (\tau + t)v$
and distance at second echo $\qquad = tv$
Now, ratio of intensity of second echo to that of first is $1 + \alpha$ and from problem 7.2 this is given by

$$1 + \alpha = \left(\frac{(\tau + t)v}{tv}\right)^4$$

or $t = \dfrac{\tau}{(1 + \alpha)^{1/4} - 1}$

(t, (b) $t = 5s$

Problem 7.4

The sound heard directly is 42×10^3 Hz as in this case there is no relative motion of the source and observer. The echo, however, will have a different frequency. The wall receives and reflects a note f' Hz where

$$f' = \frac{3.3 \times 10^2}{3.3 \times 10^2 - 10} \times 42 \times 10^3$$

Now the bat approaches this source and therefore perceives a note of frequency f'' where

$$f'' = f' \times \frac{3.3 \times 10^2 + 10}{3.3 \times 10^2}$$

It therefore hears two frequencies 42×10^3 Hz and 45×10^3 Hz.
The bat will also hear a series of beats when the two frequencies are combined and the beat frequency will be the difference between the two.

Problem 8.2

The path difference between successive rays reflected from the wing cover = $0.8 \cos (45°)$ μm = 565 nm. Rays of light of this wavelength will therefore constructively interfere at the eye and wing cover will therefore have a greenish hue.

Problem 9.2

Change in optical density at 500 nm = 0.4
applying equation 9.8 with
ϵ_{mol} = 41000; D = 0.4; x = 1 cm then
concentration of rhodopsin = 10^{-5} ml^{-1}
but, molecular weight = 40000
∴ dry weight of rhodopsin present = 0.4 mg ml^{-1}

Problem 10.3

Applying equation 10.16

$V = \Sigma RI - \Sigma E$
∴$V = I(2R_e + R_m) - E$
but from equation 10.14
$I = E/(R_v + 2 R_e + R_m)$
∴$V = -ER_v/(R_v + 2R_e + R_m)$
∴ V will only give a true reading of E when
$R_v \gg 2R_e + R_m$

Problem 10.4

(a) Applying equation 10.24 with
$E_m = -60\,\text{mV}$; E_{Na}; $+55\,\text{mV}$ and $E_K = -75.5\,\text{mV}$
then $R_{Na}/R_K = 7.4$

Now $\dfrac{1}{R_K} + \dfrac{1}{R_{Na}} = \dfrac{1}{R_m}$

where R_m is the membrane resistance ($10^3\,\Omega\,\text{cm}^2$)
$\therefore R_K = 1.14 \times 10^3\,\Omega\,\text{cm}^2$
(b) When $E_m = +40\,\text{mV}$, application of equation 10.24

gives $\dfrac{R_K}{R_{Na}} = 7.7$

If R_K is assumed constant during the action potential (i.e. at a value of $1.14 \times 10^3\,\Omega\,\text{cm}^2$), then $R_{Na} = 1.48 \times 10^2\,\Omega\,\text{cm}^2$. The equivalent membrane resistance in this case is $130\,\Omega\,\text{cm}^2$. The computed resistance is the larger because it does not take into account the fact that at the height of the action potential, the resistance of the potassium channel has already started to decrease.
(c) $1\,\mu\text{F} = 10^{-6}\,\text{CV}^{-1}$
Voltage change during action potential $= 0.1\,\text{V}$
$\therefore$ for discharge, $10^{-7}\,\text{C}$ move through per cm^2 of membrane
$\therefore$ approximately $10^{-12}\,\text{mol cm}^{-2}$ move through.

Problem 12.1

Applying equation 12.5
With $V = 600 \times 10^{-6}\,\text{V}$, $L = 1 \times 10^{-3}\,\text{m}$ and $B = 2\,\text{Wb m}^{-2}$
then $v = 300 \times 10^{-3}\,\text{m s}^{-1}$
This velocity of flow is for an area of $0.07 \times 10^{-4}\,\text{m}^2$
$\therefore$ Volume flow is $2.1 \times 10^{-6}\,\text{m}^3\,\text{s}^{-1}$

Problem 13.2

(a) 1 röntgen/hour produces $0.93 \times 10^{-13}\,\text{amps cm}^{-3}$
If the volume of the chamber is $V\,\text{cm}^3$ and the dose rate is D
then $I = 0.93 \times 10^{-13}\,VD$ amps.
(b) 1 röntgen $= 0.88$ rad.

Index